D1006237

MANEATERS and MARMOSETS

MANEATERS and MARMOSETS

Strange and Fascinating Tales from the Animal Kingdom

HEARST BOOKS • NEW YORK

ISBN 0-91-990-98-0
Library of Congress Catalogue Card Number 76-4195

Acknowledgments
Many writers contributed to *Maneaters and Marmosets,* and we are grateful to them. Specifically we wish to thank the authors of the following stories originally published in *Science Digest* magazine for permission to reprint their work here: "The African Ostrich: Bird of a Different Feather" © 1973 by George W. Frame; "Spotted Hyena: Deadlier Than Lions or Tigers" © 1973 by George W. and Lory Herbison Frame; "Those Bloodthirsty Vultures" © 1974 by George W. Frame; "Can the African Horse be Saved?" © 1972 by George W. Frame; "Giraffes—Neck-knockers of Africa © 1970 by George W. Frame; "On the Hunt with the Wild Dogs of Africa" © by George W. Frame; "Tracking and Killing Maneater Lions" © 1973 by George W. and Lory Herbison Frame; "They're Out to Save the Black Rhino" © 1971 by George W. Frame; "The Bird that Stood 10 Feet Tall" © 1966 by Roger A. Caras; "How We Found Mexico's 'Extinct' Tortoises" © 1968 by Ray Pawley; "The Horse of the Caveman" © 1968 by Bruce Frisch; "The Dragons: Past and Present" © 1969 by Daniel Cohen; "Marmosets—the World's Smallest Monkeys" © 1968 by Barbara Ford; "Are Chimps Really Animals?" © 1966 by Daniel Cohen; "Shark! Overrated Demon or Genuine Scourge?" © 1967 by William and Ellen Hartley; "Seamamm's Amazing Pinnipeds" © 1972 by Ellen R. Hartley; "The Creative Monkeys of Koshima" © 1969 by Barbara Ford; "One Animal Zoo" © 1972 by Isaac Asimov; "Leopard Cats—A New Pet Craze Gone Wild" © 1972 by Caroline T. Sullivan; "Condors" © 1973 by William W. Fox; "Meanest Animal in the World" © 1972 by Ronald Rathert; "What is the World's Deadliest Animal?" © 1970 by Gail Compton; "The Man Who Makes Pets of Gorillas" © 1969 by Dick Bothwell and David Coleman; "The Elephant Seals of Ano Nuevo and Loser's Beach" © 1975 by John M. Leighty, Jr.; "Dr. Hibben's New Mexican Ark © 1970 by Robert Gannon, by permission of Raines & Raines, Authors' Representatives; "A Watched Potto Never Grows" © 1970 by The Peabody Museum of Natural History. All rights reserved. No part of this article may be reproduced without written permission of the Peabody Museum.; "How Goes the Battle of the Whooping Crane?" © 1969 by Michael J. Walker, by permission of *American Forests* magazine; "Anglerfish—the Fisherman With a Built-in Lure" © 1970 By George Heinhold; "How to Catch a Hippo" by Robert P. Crossley from Popular Mechanics, © 1970 by The Hearst Corporation; "Life among the Gorillas" © 1964 by George B. Schaller, by permission of the University of Chicago Press.

Contents

Introduction

This is a book about animals—marmosets, maneaters, and a lot of other furry (and non-furry) creatures. We call it *Maneaters and Marmosets* because these two species lie at the opposite ends of what we might call the emotional spectrum of the animal kingdom . . . at least as perceived by human beings. Marmosets, the world's smallest monkeys, are seen by us humans as cute, adorable, and all that sweet stuff. Maneaters, in this case, lions, are seen as ferocious, mean, impolite, and other nasty things.

We cite these two opposite stereotypes to indicate that this book covers a wide spectrum of animals and animal behavior. At the same time, we hope that this book will destroy some of the stereotypes our human species applies to animals. You will discover in the following pages that marmosets have more to their character than just cuteness. And lions, even those who prefer human flesh for brunch, possess qualities other than ferociousness. You'll find that animals are complicated, just like people.

Take our story on elephant seals, for example. These beasts weigh up to three tons, spend most of the year at sea, and therefore can hardly qualify as anthropomorphic. Yet author John Leighty explains how three months every winter these giant animals congregate on small islands off the California coast where the males of the species battle each other in a series of fights of desperate violence. Why? Well, for sex, naturally. To the very few victors go hoards of females. And for the losers-—and this includes 90 percent of the male population—there's a winter of celibacy.

Upon first reading this episode, I was struck with the ruthlessness and unfairness that goes on in the animal kingdom. Thinking a little deeper, however, I realized that life among the elephant seals differed little from my own experience as an adolescent. If you're at all familiar with society in Midwestern small towns (such as the one I grew up in), you know that the men (boys) often spend a good deal of time pounding on each others' heads in order to secure a high ranking in the adolescent hierarchy. All with the aim of attracting the female of the species.

Remembering this (which was somewhat painful as I never achieved too high a place in this social order), I began to empathize more and more with those seals who took a beating and were banished to female-less shores of the mating island —appropriately named "Losers' Beach."

There's a lesson in this for all of us. Don't be so quick to stereotype animals. There's a lot in their behavior that, if you'll be honest with yourself, you should be able to identify with.

On the other hand, you probably didn't fork over good money for this book to be taught a lesson. And for those of you who feel this way, don't worry about it. Because the following pages are filled with plenty of entertaining and fascinating animal stories, which you'll enjoy even if you're not looking for an educational experience.

—DICK TERESI
Editor, Science Digest

MANEATERS and MARMOSETS

The Bird that Stood Ten Feet Tall

by Roger A. Caras

How splendid is isolation? In the case of natural history, it can be splendid indeed.

Consider the Galapagos Islands, an area with strange forms of life well known to every reader in the natural sciences. Or the Australian subcontinent, whose isolation gave enormous scope to the strange world of the marsupial and provides ecologists and taxonomists with problems for generations to come.

But one of the greatest isolation stories of all concerns the strange flightless birds of New Zealand.

New Zealand's 105,000 square miles (that's two-thirds the size of California) have apparently been remote for a very long time. Although evidence that this small group of islands was insular before and during the Mesozoic era isn't conclusive, it appears that this was so. There are no indications of a land connection and there is a marked absence of terrestrial vertebrates including dinosaurs in the fossil record.

When man came upon these, the most southern and western of the Polynesian islands, the only native mammal was a bat. Just as the ecological niches of Australia were filled by marsupial mammals, these niches in New Zealand were filled by birds, many of which could not fly. With no land predators, flight was not necessary. Evolutionary processes disposed of the unnecessary and energy-consuming function in many of the native species.

The grazing animals of ancient New Zealand were, perhaps, the most wonderful birds of all for these were the great moas. There were six or seven varieties of moas but the mightiest of all was *Dinornis maximus*. Here was a bird that stood taller than any animal alive on this planet today with the exception of the giraffe and the African elephant. The leg bones of *Dinornis* were heavier than those of our largest draft horse. A moa 10 feet tall or more consumed as much fodder in one day as our largest bullock. Its gizzard stones were as large as softballs and a single bird would carry around a bushel basket full. Here, indeed, was a wonderful bird.

The flightless ones are known as *ratites*, from the Latin word for raft *(ratis)*, referring to breast bones that are flat, unequipped with a keel that can support flight muscles.

It is not known for certain whether the moas and a number of the other flightless birds of New Zealand were descended from birds that could fly, or from birds that never had the power of flight, such as penguins. We do know that the moas were without flight for a very long time. Fossils have been found 75 feet down in the lava-baked red clay at the foot of Mount Horrible. That clay was laid down somewhere between two and seven million years ago.

No one knows when the last *Dinornis* died and we cannot know for certain if any of these 10-to-12-foot giants were seen by man, although the people who inhabited the islands of New Zealand before the coming of the Maoris in 1350 A.D. are known to history as the Moa-Hunters.

It is known that the Moa-Hunters hunted the *Megalapteryx* moas, birds no more impressive than a turkey. We do not know when these forest-dwelling moas died out. Indeed, we do not know for certain that they do not still exist in one of the many extremely remote valleys on South Island or on much smaller Stewart Island.

On a visit to New Zealand a couple of years ago, this author heard several stories about extant moas on Stewart Island.

While there was no evidence at hand that would justify such an assumption, there was a marked persistence of the stories from people who should know. The moa should not be counted out just yet. There is still some looking to be done. It is a fair assumption, however, that any moas found alive today would not be in the 10-foot-plus category of *Dinornis.*

Some moral support for the "moa-still-with-us" school is found in the rediscovery in 1948 of the flightless Notornis or takahe. Until a breeding pair and subsequently a small colony were found in remote Pyramid Valley a scant 18 years ago, the species was universally accepted as being extinct. Perhaps that does support the moa-now story, and perhaps it does not. It does show that New Zealand has secrets, however.

Impressive among New Zealand's other flightless birds were *Aptornis,* a three-foot-tall relative of the common woodhen, and *Cnemiornis,* a flightless goose. *Harpagornis,* a long-legged eagle with pitifully small wings, probably did fly to a limited extent. He was on his way to the flightless state when he inexplicably vanished, it may be assumed, forever.

Today the Notornis exists, along with the kiwi and the kakapo, to remind us of New Zealand's strange avian heritage. Now that these land-bound birds face the competition of introduced species, now that dogs, cats, and pigs roam the islands, they will probably vanish eventually as did the other species.

The kakapo is the world's only flightless parrot and had vanished for 50 years until found again seven years ago in South Island's west coast fiordland. The kakapo's status as a truly flightless bird, it should be noted, is open to question. It does have a slight keel on its breast bone and it can glide on a downhill run, or descend from a tree into which it has climbed. It is a seriously endangered species and a great deal of effort is being expended to preserve this rare and interesting creature out of the past for as long as possible.

Truly flightless and also endangered is the living symbol of the New Zealand nation and people. The kiwi is a bird so associated with this tiny South Pacific nation that New Zealanders at home and abroad are inevitably called "kiwis," and take the nickname as a compliment.

The kiwi *(Apteryx)* is a strange little nocturnal forest dweller and is the only known bird with nostrils out at the end of the beak instead of back at the base. Its feathers are like hair, its sense of smell is extremely keen, according to reports, and its eggs are larger in proportion to body size than those of any other bird, by approximately 25 percent. Overall, the kiwi is chicken size, although a little chunkier in build. It is a slightly ridiculous-looking but fascinating bird that has survived into our time by great good luck.

New Zealanders are attempting to assure its survival for as long as possible, for scientific as well as esthetic reasons. As a New Zealand government pamphlet entitled, "KIWI S.O.S." put it:

"Please give the kiwis a break. Let us keep them with us on the mainland, not merely as a few survivors on offshore islands with their replicas on coins, trade marks, and advertisements."

How Goes the Battle of the Whooping Crane?

by Michael J. Walker

Many people have lived out their lives in New York City without ever seeing a truly wild animal. And yet, at a party in the midst of this "wildlife desert," I was asked by all 10 present about the whooping crane. None had ever seen a whooper, but they were all visibly concerned with its future, asking questions such as : "How many are left this year? Why is it faced with extinction? Can we save this bird? What are we doing to help?"

I asked them why they were so interested, and got a query in return: "How can we as people survive if a harmless beautiful bird can't?"

Others expressed the fear that another member of the living world may expire, joining a growing list and raising the prospect that much of our precious wildlife heritage will be lost, leaving only rats and starlings.

Then I devoted much of my two-hour stay at the gathering discussing statistics about population, habits of the bird and actions of people that have led to its threatened existence, and some of the measures being taken to help. But the question that disturbed everyone the most went unanswered: "Can we save the whooper?" I couldn't make a prediction, given the sad situation the bird is faced with.

First, the population of this bird is so small—there are less than 60 left in the wild today—it may be beyond the rescue point.

Second, the whooper takes off each spring and fall on hazardous 2,600-mile migrations from southern Texas to northern Canada, during which time much if not all of the flock could be wiped out by careless hunters, sudden hail or snow storms, high-tension lines and low-flying planes.

Third, it is a poor breeder, most parents hatching two eggs a year but usually rearing only one chick. It also takes five years for the birds to mature before they can breed.

Fourth, many resting and feeding sites have been so disrupted by people and progress that it is doubtful a sizeable population could thrive.

But at least in the case of the whooper—as contrasted with many others, like the passenger pigeon, that expired in the past—widespread public concern has been aroused and measures are being taken.

We are, for example, in the critical stage—dubbed by the Interior Department as "Operation Whooping Crane"—of an exciting joint U.S.-Canadian program to save the bird from extinction.

Involved in Operation Whooping Crane has been the taking of extra eggs from the nests in Canada, artificially hatching the eggs at the U.S. Fish and Wildlife Service's Patuxent Research Center in Laurel, Md., and rearing a flock of whoopers there in captivity. The hope is that eventually a flock of breeders can be reared to produce stock (eggs and chicks) to bolster the whooper population.

In this joint U.S.-Canadian venture, scientists at Patuxent for several years studied the closely related sandhill crane to learn breeding and feeding habits. During the late spring of 1967 and 1968, a Canadian biologist took one whooper egg from several nests, knowing that if care were exercised the mother birds would not abandon the remaining eggs.

On five different occasions since 1966, eggs were removed

from nests in Canada and flown to the Patuxent Research Center in Maryland. Forty-one of a total of 50 eggs were hatched, and 19 of the chicks survived and are thriving, bolstering hopes that a breeding flock will develop to maturity.

Thus, counting the other captive birds retrieved from the wild with injuries and nursed back to health, there are now 23 whoopers in captivity. Adding this to the 1976 wild population of 57, we now have 80 or more whoopers alive.

The fact that we have any is amazing since it was not until 1937—when the population had shrunk to less than 25—that a safe wintering home was set aside for them in the United States. In that year, the Department of the Interior established Aransas Refuge on the Texas Gulf Coast as a permanent winter haven for waterfowl and whoopers.

If the present program to save this magnificent bird is successful, conservationists are saying that 1975 was the turning point. In an ingenious joint U.S.-Canadian venture, extra eggs were taken from nests of the wild flock in northern Canada and placed in nests of the closely related sandhill cranes at the Grays Lake National Wildlife Refuge in Idaho. The sandhills hatched most of the whooper eggs, and then reared four even though they are whiter and bigger than the sandhills' offspring. The four young whoopers then followed the sandhills to their wintering home near the Bosque del Apache National Wildlife Refuge in New Mexico.

This 800-mile journey, if it becomes established as a migration route for a new flock of whoopers, is far shorter and less hazardous than the present 2,600 mile Canada-Texas migration route of the main wild flock. When the new young whoopers mature in about five years, biologists hope their distinct mating calls, ritual dances, and plumage—which differ from those of the sandhills—will naturally lead the whoopers to seek other whoopers as mates. Whoopers cannot mate and have offspring with sandhills.

Grus Americana—the scientific name for the whooper—is the tallest of North America's native birds. It stands nearly five feet tall and has a crimson crown and black-tipped wings that spread about seven feet in flight. Except for these colorings, and dark legs and bill and facial markings, the bird is a beautiful pure white and has a ringing bugle-like call. Some whoopers may live 40 years or more.

There probably never was a large population, but up to a century ago they ranged over much of the continent from the Arctic to central Mexico and from the Rockies to the Atlantic. Their habitat, however, shrank as Americans tamed the land. Some were shot, even after they came under the international protection of the Migratory Bird Convention between Great Britain (Canada) and the United States in 1916. By the 1930s, remarkably few were left, desperately struggling to survive.

Each spring the remaining wild whoopers make a hazardous flight of 2,600 miles from the Aransas National Wildlife Refuge in Southern Texas to the region of Great Slave Lake in northern Canada. In the fall they return with a few young. They usually fly high on the migrations, sometimes even out of sight, but at other times within a careless shooter's range, though very few have been shot in recent years.

In flying north from Aransas, the cranes move across Texas, Oklahoma, Kansas, Nebraska, North and South Dakota and Saskatchewan. Their nesting area is confined to Wood Buffalo Park, a vast wilderness in northern Alberta and the adjacent Northwest Territories. One of their stop-overs in this great flight is in central Nebraska along the Platte River where the birds feed and rest. This area, however, is not protected, so we don't know how long the birds can continue to enjoy this stop-over.

In Canada, breeding pairs lay one or two buff and brown, mottled eggs in moundlike nests surrounded by water. The eggs are laid two to four days apart, and the female begins incubation after the first is laid. It takes about 34 days for the fist-sized eggs to hatch, and the new chicks are about half the size of an adult robin. Growing rapidly, the young are ready for their first flights in about 90 days, and after several weeks of practice flights are ready to accompany their parents on their long journey to the wintering grounds.

Appearing in families, singles, and small groups, the whoopers usually arrive at Aransas from October through November. Each family establishes a territory of a few hundred acres which is guarded against intrusion by other whoopers, but left open for use by waterfowl and shore and wading birds which share the same wintering ground, although concentrations occasionally occur at favored feeding sites.

Here at Aransas, the birds receive more attention than at any other place, as aerial counts are made periodically, and frequent boat patrols assure their well-being. Visitors are not permitted to enter the birds' habitat, but they can observe a few from a concealed observation tower near the territory of one family group, or from an excursion boat which carries visiting watchers along the coastal canal for closer viewing.

In late winter, restlessness signals the start of spring migration to the nesting grounds, and in late March and early April the family groups begin departing. Family groups stay together until the nesting grounds are reached, at which time the young leave their parents.

In the nesting habitat the birds are by no means secure. A severe windstorm or unseasonal blizzard or hailstorm could destroy them or their nests. It is for this reason that many wildlifers believe the survival of the species rests on breeding captive birds to rear young for bolstering the wild flock each year and also creating different flocks in other areas.

One of the leading private contributors to the welfare of the birds has been the National Audubon Society, which was instrumental in getting oil companies to curtail planned operations that could have imperiled much of the cranes' winter habitat. The society initially employed a warden to patrol 7,000 acres of Matagorda Island, Texas, which some of the birds use, and later hired biologists to study the habits of the bird. Continued protection of this island is essential for the birds' welfare.

But other groups, besides the U.S. and Canadian Governments, including the National Wildlife Federation, have done much for the bird by educating the public to its plight. We still have an occasional careless hunter, as we did in the fall of 1967 in Texas when a whooper was shot near Aransas, but many sportsmen have learned to distinguish the whooper from other white birds on the basis of much greater size, long neck, trailing legs and slow wing beat.

The view of the scientist credited by conservationists as one of the leaders of rescue efforts—Dr. Ray Erickson at the Patuxent Research Center is: "The success of these and other efforts remains to be seen. We may be turning the corner because of the encouraging 1975 transplanting program in southeastern Idaho, and the prospects of reproduction by seven pairs of confined whoopers at Patuxent."

What is the World's Deadliest Animal?

by Gail Compton

It's the sea wasp, a lovely, delicate jellyfish, known technically as *Chironex fleckeri*.

"Only a bullet kills faster than a sea wasp," says one Australian scientist.

Victims usually are dead within five minutes of brushing against the sea wasp's trailing tentacles. Most of them die irrational and screaming in agony. Others are luckier. The sting of the tentacles knocks them unconscious immediately. One sea wasp killed an 11-year-old girl in 30 seconds.

Sea wasp tentacles are loaded with hundreds of thousands of microscopic stinging cells which inject a cobra-like venom. There is no known antidote.

"The venom produced by the deadly sea wasp, *Chironex fleckeri*, may rank among the most toxic substances known," says Bruce Halstead, director of the World Life Research Institute and one of the world's outstanding authorities on poisonous and venomous marine animals.

The jellyfish is virtually an invisible killer. The transparent bell-shaped body is colorless and is almost 95 percent water. The tentacles are usually tinged purple or light blue. Only careful eyes can spot it drifting in the shallow tropical waters of sea shores and beaches. A lash of burning pain usually is the first knowledge a swimmer or bather has that the jellyfish is present.

For many years scientists were baffled by quick, violent deaths that occurred with alarming frequency in waters off northern Australia. *Chironex fleckeri* finally was identified as the probable villain.

Dr. R. V. Southcott of Australia, recognized authority on the sea wasp, reported: "Death may ensue from under one-half minute to up to two or three hours, but usually takes under fifteen minutes, often in three to four minutes after contact. In many cases the victim dies before or soon after struggling the few yards to shore."

Dr. Southcott believes many deaths attributed to cramps, collapse, overeating and drowning quite probably were the result of sea wasp stings or of similar jellyfish.

Sea wasps range from the size of a child's fist to gelatinous blobs eight inches across and four or five inches deep. Trailing from the four corners of the bell are the venomous tentacles. These sometimes reach four feet in length. A single sea wasp may have up to 50 tentacles. And just one tentacle may have as many as 750,000 individual stinging cells capable of injecting venom. The trailing tentacles form a net of death for any creature brushing against them.

Basic purpose of the deadly tentacles is to secure food for the jellyfish, such as small fishes. They also serve us a defense mechanism. Man is merely an innocent victim.

Most deaths from sea wasp stings have occurred in waters around Australia where the species *Chironex fleckeri* is abundant. Australians fear this animal more than the huge sharks for which their waters are noted.

Brisbane newspapers reported 60 persons killed in the last 25 years by jellyfish in water around Queensland. In the same period of time and area, sharks claimed only 13 victims.

Species of the sea wasp are also found in United States waters off the Atlantic coast from North Carolina south to the Florida Keys. They live in the seas around the Bahamas and all the way down to Brazil. They are also found in the Philippines, off the coast of West Africa and in the Indian Ocean.

The 11-year-old girl was killed while wading in about two feet of water, just 50 feet from shore, near North Queensland, Australia. Her legs brushed against the unseen tentacles of a sea wasp. She was dead in 30 seconds.

Results of the autopsy were reported by Dr. Southcott. The autopsy revealed lungs and air passages filled with large quantities of frothy mucus. Abdominal organs, kidneys and brain were congested but otherwise normal. The heart was normal. Skin on the legs showed linear reddish brown streaks about six millimeters in width where the jellyfish tentacles adhered. Shock and pulmonary edema were listed as the direct cause of death.

Dr. Southcott also reported autopsy results on a 38-year-old male who died a few minutes after being stung by a sea wasp. Body conditions were almost identical with those of the little girl. But there also was evidence of minute hemorrhages in the brain.

Strips of skin lifted from both victims disclosed numerous stinging cells from *Chironex fleckeri* still imbedded in the tissue, Southcott reported.

Symptoms of the deadly sting are unmistakable. Pain is so intense and burning, victims scream and become irrational. Many are rendered unconscious instantly. Painful muscle spasms, rapid weak pulse, prostration, pulmonary edema, vasomotor and respiratory failure frequently result. An important cause of death is primary shock followed by drowning before rescue attempts can be made.

Chironex fleckeri is the deadliest member of a virulent group of sea animals, Coelenterata. Most coelenterates are equipped with venomous stinging cells. No other animal or group of animals has this lethal equipment.

Most people are familiar with this family of animals. Members include the pretty, purple Portuguese man-of-war, *Physalia physalia*. Drifting into southern beaches during the height of the tourist season, fleets of man-of-war become a public health menace, stinging thousands of bathers. Other coelenterates are the flower-like sea anemones and the brilliant corals which can be dangerous to scuba divers, and there are thousands of kinds of jellyfish ranging from the microscopic to monsters seven feet across with tentacles 120 feet long.

Coelenterates are primitive animals. Some, like the corals, are fixed to the ocean floor and have no means of locomotion. Others drift at the mercy of wind and tide, and at best, they are very poor swimmers. They seem defenseless and harmless. Many are brilliantly colored and attractive.

Nature compensated for their weakness by providing them with a highly sophisticated death-dealing mechanism. Stinging cells, called nematocysts, lie just below the skin surface of the tentacles. These venomous weapons make coelenterates among the most dangerous animals in the world.

The nematocyst is a remarkable apparatus. It is nature's version of a hypodermic needle. The microscopic cell consists of a capsule filled with venom. Tightly coiled within the capsule is a long, hollow thread-like tube. Pointed at the tip like a dart, the tube also has barbs to hold it fast in the victim. A tiny hair-like trigger protrudes from the capsule.

When the trigger is touched, the capsule explodes into action. Like a harpoon, the thread tube fires out, unwinding as it goes. Its length can be several hundred times the diameter of the capsule. The needle-sharp tip hits the victim and penetrates. Barbs hold it fast. Venom pumps through the hollow tube into the prey.

The explosive force of the cell discharge can shoot the venomous tube through heavy surgical gloves. Penetration of human skin presents little difficulty. The nematocyst stays imbedded and continues to inject its venom.

Tentacles of the man-of-war retain their stinging power long after the animal has drifted onto the beach, died and the body evaporated. In the laboratory, dried tentacles have retained their toxic qualities for two years.

Dr. Halstead reports a case of a man who poisoned an enemy

by feeding him soup prepared from the dried tentacles of a man-of-war.

Nobody knows how many humans suffer stings each year from the man-of-war, jellyfish, anemones and corals. Records aren't kept. Only the most serious stings are reported. Estimates by scientists range from "thousands" to "hundreds of thousands" annually.

In one day, on Miami Beach, more than 400 perople were treated for man-of-war stings. Life guards cleared and closed the beaches. This is not an unusual occurrence. Man-of-war stings have been fatal in several known cases.

In March, 1964, a man staggered ashore and died at Miami Beach with tentacles from a man-of-war clinging to his chest, arms and legs.

A thimbleful of venom from the man-of-war will kill 1,000 mice, says Dr. Charles Lane of the Institute of Marine Science at the University of Miami. The poison is almost identical to cobra venom.

From reported cases of jellyfish stings, marine scientists set the fatality rate at roughly nine percent. Out of 502 cases there were 46 fatalities, according to figures presented by Dr. Halstead. How many deaths listed as drowning were the result of jellyfish stings can only be guessed.

Some recent figures are given by Dr. John Barnes of Cairns, Australia, a clinical expert on jellyfish stings. Reporting for his area, Dr. Barnes said that from November 1956 to May 1960, there were 116 known victims stung by jellyfish just in the vicinity of Cairns.

Other reports from various parts of the world merely list the number of people stung as "many," "several" or "hundreds."

Effects can range from a mild dermatitis to death. Some people are more susceptible than others to jellyfish stings. As with bee stings, this sensitivity can result in death if stings are repeated over a period of time.

"Severity of a sting depends on how long the victim is in

contact with the tentacle and how many nematocysts are encountered in that contact," Dr. Halstead said.

There are three steps to take if you are stung by a jellyfish. Relieve the pain. Alleviate the neurotoxic effect of the venom. Control primary shock.

For the man at the beach this means removing the tentacles from the skin immediately. Use sand, gloves, a gunny sack, sea weed or anything handy. Don't touch the tentacles with bare hands.

Then apply alcohol, diluted ammonia, sun tan lotion, oil or any other similar lotion at hand. These will help stop the nematocysts from continuing to discharge venom into the skin.

Cardiac and respiratory stimulants should be administered if such failures occur. Oral antihistamines and cortisone have proved helpful.

Just remember, be careful where you swim, especially in warm or tropical waters. If you are stung, get out of the water immediately. Try to spot what kind of jellyfish it was; it will help in the treatment later.

We do not have adequate methods for treating jellyfish stings. No antidotes. Just pray it wasn't the lovely, delicate *Chironex fleckeri* that brushed against your legs. If you have enough time.

Giraffes: Neck-knockers of Africa

by George W. Frame

Giraffes are normally considered benign creatures, but they're really some of the most spectacular fighters in Africa. Their method of fighting is best described as sparring, or better yet, necking, and is used by the bulls to determine their rank in the herd. And it's a sight to behold.

The combatants stand side by side, either facing in the same or opposite direction. They never try to bite, nor do they strike each other with their legs as they would do to a predator. Instead they stand still and swing their long necks in wide arcs, each trying to strike his horns against his opponent's head, neck, body, or legs.

The opponent attempts to dodge the blow so that it will be either glancing or a complete miss. Then the one who dodged takes his turn at swinging. Usually the hits are not too severe and the swinging of their necks is graceful rather than vicious. Between bouts both giraffes sometimes stand and rub their necks together. This necking often then results in sexual play.

But this strange ritual may one day come to an end because these lofty animals are rapidly declining in number and in range. Actually, this decline has been going on for the past 6000 years. As climatic changes created the Sahara desert, giraffes declined in numbers, and those which survived in isolated areas were exterminated by hunters. Eventually, the giraffe became extinct in all of North Africa. The many excellent rock paintings of giraffe at Tassili in the Sahara, are testimony to these former days.

Pictures on ancient Egyptian tombs show Nubian slaves carrying bundles of giraffe tails and evidently the giraffes may have been killed for this prize alone. Perhaps these tough tail hairs were used for sewing or stringing jewelry.

European settlers in Africa during the 1800's hunted the camelopard (giraffe) for its meat and skin, which was used to make leather harnesses and whips. Africans, too, killed the giraffe for meat, being especially fond of the marrow in the large leg bones. The strong sinews served for bowstrings and for the strings of musical instruments. Tails were sold for a very nice profit. One man even killed giraffes regularly in the area now constituting Kruger National Park, for the purpose of selling the bones as manure. Wagonloads of giraffe hides and bones were a common sight as late as the 1870's.

Particularly during the past 30 years, man has accelerated the decline of this creature. Expanding agricultural, industrial and residential development have severely reduced the giraffe's habitat and extensive poaching has hastened their demise. Today, the giraffe survives on less than ten percent of the African continent.

Poaching in national parks and game reserves is a never-ending problem. In the Kidepo Valley, Uganda, cattle raiding gangs from the southern Sudan often kill giraffe and buffalo. The local poachers showed great ingenuity by wearing square shoes to confuse the rangers who tried to track them. Once the park warden found footprints leading away from the park, but when he tracked them he found that the poachers had actually walked into the park backwards! The poachers were later followed and arrested.

Though giraffes don't stand much of a chance against poachers, they're well-equipped to fend off predators. Even when fighting among themselves they've been known to do serious damage. In such fights they seldom stand still except when they spread their legs far apart to get good leverage. At these times the blows they deliver with their swinging necks are so strong that I have sometimes heard them nearly 100 yards away. And I've seen a good hit push a giraffe sideways for several feet.

Although their necks and bodies are covered by inch-thick skin, they often raise bumps and swellings on each other. In Kruger National Park (South Africa) during one such fight, a bull was knocked unconscious for 20 minutes.

Another time in the same park, tourists watched as one giraffe killed another large bull. Examining the body, the warden found a horn hole behind one ear—the first vertebra was fractured and bone splinters had pierced the spinal cord.

The same ferocity is used against predators, only in this case kicking is also used. And a giraffe's kicking power is awesome. More than once in South Africa, a giraffe has walked up to a car, turned around, and kicked in the radiator. In Uganda, a giraffe struck a vehicle with its forelegs, smashing the headlights and windshield and bending the steering wheel.

Lions are the major predator of giraffe. In attacking an adult giraffe they usually jump for the neck, and grab hold with their teeth and claws until the giraffe falls. But often this backfires. There are several records from East Africa of a dying giraffe falling on an attacking lion and crushing it to death.

A game warden in South Africa once saw a lion chase a giraffe for some distance before catching it. The lion leaped on the giraffe's rump, but slipped off and received kicks from both hind feet of the giraffe. The lion's chest was crushed and all of its ribs were broken.

Sometimes, though, the giraffe is not so lucky. Recently, in Kenya's Tsavo National Park, a group of tourists watched a pride of five lions follow a giraffe as she was about to give birth. As soon as the baby giraffe was born the lions devoured it in full view of the startled tourists.

The leopard is another, but less important, predator. Near

Nairobi a male giraffe was about to browse on a tree which contained a sleeping leopard. The leopard sprang on the giraffe and mauled its neck until it died.

There are also at least three records in Kenya of a crocodile pulling an adult giraffe into the river and drowning it.

Even more unusual is the report of a giraffe found dead with a dead python under it. Apparently the python coiled itself around the giraffe's neck and strangled it to death. Then the giraffe's great weight pinned the python against the ground, crushing it to death.

Giraffes often depend on their speed for protection, being able to gallop at 35 miles per hour over short distances, and able to maintain 28 to 32 mph for several miles. When galloping, they twist their tail over their back like a corkscrew. They can almost outrun horses, but because giraffes are short-winded, horses can eventually overtake them. The giraffe's lung capacity of only about 13 quarts of air, compared to 33 quarts for the horse, explains why it lacks stamina.

Camouflage is another protection. A giraffe's spot pattern remains constant from birth throughout life, though colors tend to darken with age. Researchers thus identify and keep track of individual giraffes by keeping a photographic record of the spot pattern on their necks. The camouflage effect of these spots and colors is incredible. More than once I've observed a giraffe in the bush about 30 yards away for several minutes before I realized I wasn't watching one animal but a whole group.

Giraffes usually associate in small groups of three to 15 individuals, but more than 100 have been seen together. Dr. J. B. Foster studied giraffes in Nairobi National Park and found that the herd structure is variable. Large herds were never composed of the same individuals in two consecutive observations and some males formed bachelor herds and tended to live in more forested areas. Young males joined these bachelor herds when they reached three years of age, and they apparently stayed in these groups until fully adult at about seven years.

Giraffes drink by straddling their forelegs sideways or by

bending their front knees forward until their head can reach the water. Drinking usually lasts about 20 seconds, and then they quickly straighten up again. They may do this five or six times, possibly because it is fatiguing to remain crouched for too long at a time or because they fear being surprised by predators.

The physiology of giraffes has long intrigued scientists, because their long necks suggest special adaptations. Several years ago three Americans measured the blood pressure of two wild giraffes in Kenya by immobilizing them and implanting sensors in their arteries. They found that the blood pressure at the base of the brain of a standing giraffe is about the same as in man. However, when the giraffe lies with its head on the ground, the blood pressure becomes more than twice as great. The arrangement of arteries in the giraffe's head is such that when it lowers its head the blood, instead of congesting in fine vessels, overflows into a larger artery.

In South Africa male giraffes have been recorded up to 19 feet tall and weigh up to 2600 pounds. Full-grown males in East Africa sometimes exceed 17 feet. This lofty height occasionally clashes with civilization. When the railroad was first built through Kenya and Uganda, giraffe a number of times walked into and broke the telegraph line. The only solution was to raise the height of the wire.

Newborn giraffe are nearly six feet tall and weigh about 120 pounds. Calves are born singly at any time of the year, after a gestation period of about 15 months.

But just as the bison has disappeared from nearly all of North America, the giraffe will disappear from most of Africa. However, many African nations have established national parks and nature reserves where the giraffe can survive in limited numbers. In the Transvaal of South Africa, for example, hunters reduced the giraffe population to only a few dozen in 1902, but with strict protection they have since multiplied to nearly 3000. Thanks to the concern of conservationists all over the world, the African giraffe does have a future—although its numbers will be greatly limited.

Anglerfish: The Fisherman with a Built-in Lure

by George Heinold

Past experiences forewarned me that the half-dozen teen-agers hurrying excitedly into my yard believed they had encountered a mysterious member of the animal kingdom. Two of the boys were lugging between them a pole from which sagged a bulky creature wrapped in the folds of a minnow seine net. The procession reminded me of bearers in a Frank Buck movie bringing back alive some trussed beast.

"We've captured a sea monster!" one of the bearers cried. "We found it creeping over a rock at the seashore. It's a horrible looking thing with a mouth and teeth bigger than a shark's. No one down there knows what it is. Do you?"

Because I am known as an advanced amateur naturalist, people often beat a path to my door with various specimens of the wild kingdom they suspect are dangerous. They seldom are, but I couldn't blame the lads, for the specimen they had in tow did resemble a bona fide sea monster—at least a mini version of one. Their yard-long captive was an almost fully-grown angler, a fish that dwells in the shallow northeastern coastal waters from Newfoundland to North Carolina. The angler is a fish whose grotesque features and strange habits have given him many aliases. Among the more common are allmouth, goosefish, fishing frog, molligut and because of the formidable size of his mouth, Maine fishermen call him lawyerfish.

The angler is a fish which, if it could swim back into the primitive seas of the Devonian period, would not look out of place. One's first impression is that he's all head; the second, that the head is all mouth. The angler's head makes up nearly one-half of the entire fish. Comet-shaped, the rest of the angler's body tapers so sharply to the tail that the nearest of the common creatures he resembles would be a giant tadpole.

Only in shape does an angler resemble a tadpole. In adulthood his huge mouth grows to more than a foot in width and is directed upward, with the lower jaw thrusting beyond the upper. Both jaws contain numerous teeth resembling needles permanently exposed, a dental arrangement which causes the angler to grin at the rest of the nautical world as macabrely as the skull of a skeleton. His yellow and black eyes are at the top of his head and, like the mouth, point upward.

The angler is different from other fish we know in various ways. Lacking gill covers, he has instead small openings behind the pectoral fins on the forward end of his body. Through these he draws the water from which he extracts oxygen as it bathes his gills. The angler doesn't have scales on his deep brown to blackish body. Instead, rows of fleshy tabs cover his head and body.

Above the mouth of the angler is a slender elongated spine, or what would be the modified first ray of a dorsal fin in other fish. It has a tufted end not at all unlike one of Izaak Walton's finest lures in appearance. The angler's fishing technique is even more effective than Izaak's. Lying still with mouth shut, the angler daintily waves the tufted end which is his "lure" with the tentacle which is his "rod." Impelled by a desire for food or by curiosity, fish of all kinds come to inspect the lure and are seized and drawn into the opened mouth, the teeth of which, like the bars of a jail cell, effectively prevent escapes.

Classified by ichthyologists as *Lophius piscatorius*, anglers have been known to attain lengths of four feet and weights of 70 pounds. There are at least a dozen known species of his family dwelling in the Atlantic, Pacific and Indian Oceans. Of these

Lophius is the largest, as well as the one encountered by seagoing man most often, due mainly to the fact that the angler *Lophius* fishes where man fishes.

Other species of anglers, fishing in deeper waters, are less known to the average seaman. These are called deep-sea anglers. Most of the deep-sea anglers are much smaller, but all have physical characteristics of the *Lophius,* including rod and lure. They also have some features all their own.

One of the deep-sea anglers has on the end of his fishing-rod a bulb which emits a purplish light. Scientists believe that this can be turned on and off at will. Like the glittering chromes which cover lures made by man, this luminous organ attracts victims in the dark depths of the seas.

In deep-sea angler society, the larger and more robust females perform all the necessary functions of life but one. Puny males, often only an inch long, attach themselves to any part of a female's body and are eventually completely fused to it, becoming mere appendages on their wives. The males' bloodstream connects with those of the females', and in this way they receive their nourishment. For the rest of their days their only function is fertilizing the eggs when spawning time arrives.

The *Lophius* angler's method of reproducing its progeny, on the other hand, resembles more closely that of salmon, bass and other fish. I once had an opportunity to see this for myself one June morning when I accompanied a lobsterman tending his pots off Cape Cod.

What first drew my attention to the spawning rites were heavy splashes. Then I saw the huge head of an angler emerge above the surface like some monstrosity from a science fiction story. The head ducked and reappeared again and then again, carrying on like a playful porpoise swimming in a circle. Never before had I seen an angler move with such vim and vigor.

"That's a molligut spawning," the lobsterman told me, with evident disapproval. "I don't rightly know if that's a she releasing her eggs or a he fertilizing 'em with milt. All I know is that

I'm going to pull up all the pots I've set around this reef—and fast!"

"Why?"

"Because where one pair of molliguts spawn others will follow. In a day or so every bit of water here will be messed up with the sticky jelly veils the eggs hatch. I want no part of that!"

From what I saw a few days later when, alone this time, I sailed out to the reef (and also from consulting authorities), the veils in which the angler's eggs incubate and hatch are transparent mucus films from four to five feet wide and up to 30 feet in length. They are buoyant, undulating like gelatinous rafts. As hatching time approaches, a matter of four or five days, the hundreds of thousands of eggs they contain cause them to turn to the color which has given them the common name "purple veils." After hatching, the multitudes of young anglers float on the surface of the sea until they attain a length of one and a quarter inches, becoming prey which helps sustain other fish and birds. The survivors then take up a life among floating seaweed, remaining there until, near summer's end, they develop enough to sink to bottom.

One day I watched a haddock fisherman in Maine land a large angler whose belly bulged like an over-inflated balloon. "Watch this," he urged, after turning the angler on its back and producing a knife. "You might find it interesting."

When the angler's mottled white stomach was opened, seven haddock averaging 16 inches in length were revealed. All were in good condition.

"I always do this when I land a big lawyerfish with a full belly," the fisherman told me. "It's worth the trouble. They have slow digestions, and the fish I take from 'em are almost always in fit condition to sell. I've opened laywers with as much as 27 herrings and 33 mackerel in 'em."

"Do you know if other fish eat the larger lawyerfish?"

"Once in awhile they find one in the gullet of a shark. Never heard of 'em being found in other fish."

Further exploratory surgery of my own on three large anglers revealed that they are indeed excellent fishermen. Their stomachs contained approximately two bushels of flounders, whiting, butterfish, porgies, eels, lobsters, crabs, blackfish, small rays and one beer can. The last angler I examined in this way was the one the boys I mentioned earlier brought to me for identification. Along with porgies and butterfish, the gullet of this one also held a bluebill duck and two terns.

The omnivorous angler can also be regarded as a hunter. This accomplishment is borne out by the name goosefish, one of his most widely-used aliases. Sea-birds that dive for food have often been found inside of him. The late naturalist, William T. Hornaday, cites in his "The American Natural History," a case of an angler with seven ducks inside him.

Just how the angler attracts diving-birds to within reach of his jaws is a question students of marine life debate. Does he do it by hoaxing them into thinking that the waving tentacle that is his lure is a tidbit, a worm perhaps? So far as the angler *Lophius* is concerned, I believe that is entirely possible. It is during the colder months of spring and autumn, that he moves into the shallower waters, the feeding grounds of birds.

Hornaday and his colleague, Dr. G. Brown Goode, were among the earlier naturalists who classified the angler among the order of what they referred to as "foot-fishes." This would suggest that anglers, like the batfish, also of sinister appearance, are able to walk the ocean floor. There is some truth in that, but only to a limited extent. Batfish have strong and broad rear pectoral fins, as well as an additional pair of fins ahead of these. Anglers do not have the latter. Batfish actually walk over the bottoms of the seas with a four-footed gait and can also execute little hops. Anglers cannot do these things.

Nevertheless, the angler can creep slowly on bottom when compelled to. He presses his wide pectoral fins firmly on solid surfaces, working them in unison. He does this when he moves in close to shore during the spring and autumn months to appease his insatiable appetite. Several which have been brought to me, including the duck-eater captured by the boys I have mentioned, were found on submerged or above-water rocks or ledges. The angler can stay alive out of water considerably longer than most fish, and one that was landed on a boat I was on was still alive after three hours. This one was kept by the fisherman to be hickory-smoked and used as food. Though historical documents indicate that the angler was used as food by coastal Indian tribes, this was the first and only time I have ever known a modern man to eat one.

I do not mean to imply, however, that the anglers which are found on rocks and ledges deliberately climbed out of their element. The evidence indicates that they had set up ambuscades among inshore rocky places, spots where forage fish abound, were caught high and dry by rapidly-falling tides, and were discovered before they could inch their way back into water.

Fortunately for seaside resorts catering to summer vacationists, the angler *Lophius* is considerate enough not to be conspicuous during the height of the tourist season. Goblin-like as he is, the angler much prefers the Halloween climate.

On the Hunt with the Wild Dogs of Africa

by George W. Frame

The Land Rover bounced across the dusty plain inside Ngorongoro Crater, as game biologist John Goddard struggled to keep control of the wheel. About 50 yards to the left of us, and slightly ahead, a pack of nine African wild dogs chased close behind a fleeing gazelle.

The wild dogs ignored us as they concentrated on catching their prey. We were doing about 30 mph, but barely keeping up with them as John stepped on the gas and swerved wildly to miss the holes and ruts in the plain.

After more than a mile of running, the lead dog put on a final burst of speed and came alongside the Thomson's gazelle. With a quick slash of his teeth, he tore open the gazelle's abdominal wall, and the intestines began to protrude. Another quick slash of the teeth, and most of the viscera came tumbling out, forcing the gazelle to stop running. In deep shock and apparently oblivious to pain, the prey stood still as the three leading dogs began to tear their way further into the abdominal cavity. Within minutes the gazelle was dead.

Cruel? The casual observer may think so. For this reason and because their chases are so successful, the African wild dogs, more properly called Cape hunting dogs *(Lycaon pictus)*, have received a notorious reputation of being deliberately malicious killers, more ruthless than any other African predator. Up until

1958 in Rhodesia, for example, wild dogs were being shot and poisoned. And in South Africa's Kruger National Park as recently as 1960, wild dogs were shot on sight as part of a general policy on predator reduction.

In the decade of the 1960s, zoologists began to turn their attention to predators of the African continent. Cape hunting dogs were among the most misunderstood, so it was with enthusiasm that I accepted the offer of Tanzania's Ngorongoro Conservation Area to assist John Goddard in his zoological studies.

Cape hunting dogs inhabit most of Africa south of the Sahara, but they do not occur in jungle habitat. Basically, wild dogs are animals of the open plains.

They are not really closely related to domestic dogs, but are a four-toed relative of the wolf and fox. Their large, rounded, almost bare ears stand starkly erect. A thick ruff of fur covers the neck and throat, but the rest of the body has only short hair. Their pelage exhibits endless variations of black, white and yellow splotches. Muzzle color is always black, and the tip of the tail is almost always white.

The average weight of wild dogs in East Africa probably does not exceed 40 pounds, but members of the southern race are larger. Those found in Kruger National Park, South Africa, weigh from 50 to 60 pounds and are several inches taller at the shoulder.

A strong musky odor is secreted by the dogs. Whenever I examined the grass where the pack had been sleeping, I was always impressed by the intensity of the odor. Probably it plays an important part in recognition of pack members, and it may be of value in helping the pack stay together when chasing prey.

Cape hunting dogs have three very distinct cries: a combined short deep bark and growl when frightened; a chattering or high-pitched twittering when excited; and a soft, musical, bell-like "hoo" when members of the pack become separated. The last sound can be heard at a distance of more than two miles. Less common vocalizations are whining and yelping.

I find the musical "hoo" especially interesting, because of its uniqueness. Once when John and I were unsuccessful in keeping up with the pack, we fell far behind with one lone straggler from the group. The rest of the dogs were nearly two miles away, and out of sight because of some high vegetation in between. The lone dog was clearly lost, and appeared uncertain about which direction to go. All of a sudden his ears perked up, and after several seconds of pausing he took off running in the direction of the distant pack, apparently seeking the source of the bell-like "hoo."

As the straggler approached the pack, some dogs adopted the stalking posture in which the head and neck were held horizontally, the shoulders and back hunched and the tail hanging. When the lone dog rejoined the others, all members partook of the greeting ceremony. This mainly consisted of face-licking and poking the nose into the corner of the mouth. Usually the begging dog crouches with chest close to the ground, and rump, tail and head pointed stiffly upward.

The primary functions of the free-ranging wild dog pack are hunting and the protection and rearing of litters. According to zoo records, the gestation period is about 69 to 72 days. Litter size varies from two to as many as 12, with the average being about seven pups.

When a wild dog is ready to bear pups, she seeks a burrow abandoned by aardvarks, warthogs or hyenas in which to den. During this time the rest of the pack continues to sleep out in the open on the grass, as they always do.

When the pack leaves the vicinity of the den to hunt, they leave behind at least one member, either male or female, to guard the pups. Within minutes after devouring the catch, they return to the burrow and disgorge meat in front of the begging pups and the guards who remained behind.

At three or four months of age, the pups begin to lie out in the open, and to accompany the pack on its hunts. Naturally, the pups lag far behind in the chase, but the adults guard the carcass and wait for the pups to catch up and stuff themselves, before they too join in.

At hunting time, which is usually during the day, one dog arises first and arouses the others by nudging them with its nose. For about 10 minutes they actively play and chase each other, making twittering sounds in the process. Play is sometimes followed by running after the first suitable game, or by a careful stalk followed by the chase the moment the prey becomes aware of them.

Generally, the pack has a regular leader, who takes the initiative in selecting and catching the prey. He is supported by one or two others that often stay about 100 yards behind. They usually attack in a column, the following dogs intercepting the prey if it tries to escape by zig-zagging or by doubling back.

Chases vary in distance from as little as a few yards to as much as three miles. Speeds of 20 to 35 mph are reached. Frequently, the dogs become separated, with the slower ones lagging as much as a half mile behind. The one or two leading dogs typically catch their prey by snapping at the abdomen and disemboweling the animal. As the lead dogs begin to devour the prey, the stragglers arrive, and they too partake of the feast. Behavior of the feeding dogs is quite friendly, and each one seems to try to outdo the other with submissiveness. Usually the head, hoofs, some fur and bloodstained grass are all that remain 20 minutes after the kill.

Because Cape hunting dogs prey upon the most abundant species, their diet varies considerably throughout eastern and southern Africa. In Ngorongoro Crater, the most sought-after year-around prey is Thomson's gazelle, with juvenile wildebeest and Grant's gazelle ranking as poor second and third choices. During the month of January, at the height of the wildebeest calving season, the dogs specialize in catching the calves. There are at least a dozen other species that are prey for the dogs.

Hyenas are very abundant in Ngorongoro Crater. Whenever the wild dogs make a kill within the boundaries of a hyena clan's territory, they are certain to be confronted by the resident hyenas. On more than one hunt I have seen several hyenas steal the prey from the one or two lead dogs. But when the rest of the pack arrived at the kill, they were sometimes able to drive away the hyenas by snapping at their rumps.

I could not help but laugh as two of the dogs snapped at the rear of a hyena. The poor hyena responded by sitting down, but pulling itself along with its front legs. While dragging its behind on the ground, it snapped and growled over its shoulder at the dogs. However, on the whole, the hyena was decidedly timid and beat a hasty retreat from the dogs' attack.

There are recorded cases of wild dog packs chasing leopards and even adult lions in a seemingly determined manner, but the prey escaped the dogs by climbing trees. In the Kafue National Park of South Africa, wild dogs have twice been seen killing and eating old lionesses. But, there are no records of their ever having killed a hyena.

Predators other than hyenas capitalize on wild dog predation. U. De V. Pienaar, working in South Africa's Kruger National Park, mentions an instance in which a pack of eight wild dogs were hunting. They were followed by five vultures flying low and directly behind the pack. Behind the vultures, a hyena with two half-grown pups followed. The hyenas in turn were followed by four saddle-backed jackals. Successions of predators like this have often been seen in Ngorongoro Crater, too.

Dependence of the dogs upon the pack for hunting became startlingly apparent to me during one of my observations in Ngorongoro Crater: The pack of nine dogs chased an adult male Grant's gazelle for three miles. During the chase the dogs became widely separated, with one dog more than half a mile ahead of the others. He closed to within five yards of the gazelle before realizing that he was alone. His reaction was an abrupt stop, and the gazelle stopped, too. The dog looked back, looked at the gazelle and then turned around and hurried back looking for the other pack members. The exhausted gazelle, which had been easily within reach of the dog, responded by running off. This was a case where there was no question in my mind as to pack importance. The dog abandoned his prey rather than attack it alone.

Perhaps it was in Kruger National Park, in 1947, that the first general awareness developed of the predator's role in population regulation. Park warden Stevenson-Hamilton reported that destruction of the park's formerly large wild dog packs coincided with a tremendous increase of impala. In 1969, Dr. Pienaar suggested that increased numbers of wild dogs would actually benefit the impala population, which is currently so abundant that the vegetation of the park is beginning to suffer. This is further supported by the fact that they tend to select the most abundant species as their prey.

The data that Richard Estes and John Goddard published in the January 1967 issue of *Journal of Wildlife Management* reemphasize the importance of African wild dogs as population controls. Based on 50 observed kills within Ngorongoro Crater, they concluded that during the wildebeest calving season the dogs specialize upon the abundant supply of new calves. Throughout the rest of the year about two-thirds of their prey consists of Thomson's gazelles, most of which were adult territorial males. Such a high percentage of territorial males captured by the dogs exemplifies how predation benefits the prey species: Normally in gregarious, territorial antelope species there is an excess of healthy adult males which are unable to reproduce because of lack of suitable territories. Removal of territorial males by predation opens up new territories for younger, more vigorous males.

Modern wildlife management concepts have terminated the policy of indiscriminate destruction of Cape hunting dogs and other predators in African national parks and game reserves. Those well-meaning individuals who persist in killing wild dogs are becoming a minority, and the future of the species appears less precarious. The Cape hunting dog is both an interesting and important predator and it deserves protection.

One Animal Zoo

by Isaac Asimov

In 1880, a stuffed animal arrived in England from the newly discovered continent of Australia.

The continent had already been the source of plants and animals never seen before—but this one was ridiculous. It was nearly two feet long, and had a dense coating of hair. It also had a flat rubbery bill, webbed feet, a broad flat tail, and a spur on each hind ankle that was clearly intended to secrete poison. What's more, under the tail was a single opening.

Zoologists stared at the thing in disbelief. Hair like a mammal! Bill and feet like an aquatic bird! Poison spurs like a snake! A single opening in the rear as though it laid eggs!

There was an explosion of anger. The thing was a hoax. Some unfunny jokester in Australia had stitched together parts of widely different creatures and was intent on making fools of innocent zoologists in England.

Yet there were no signs of artificial joining. Was it or was it not a hoax? And if it wasn't a hoax, was it a mammal with reptilian characteristics, or a reptile with mammalian characteristics, or was it partly bird, or *what?*

Slowly zoologists admitted that the creature was real, however upsetting it might be to zoological notions. There were increasingly reliable reports from people in Australia who caught glimpses of the creature alive.

But disputes over the nature of the duckbill platypus, now known scientifically as *Ornithorhynchus anatinus*, went on heatedly for decades.

When specimens were received in good enough condition to allow study of the internal organs, it appeared that the heart was just like those of mammals and not at all like those of reptiles. The egg-forming machinery in the female, however, was not at all like those of mammals, but like those of birds or reptiles. It really seemed to be an egg-layer.

It wasn't until 1884, however, that the actual eggs laid by a creature with hair were found. And not only those of a platypus, but also of another Australian species, the spiny anteater. The find merited an announcement. A group of British scientists were meeting in Montreal at the time, and the egg-discoverer, W. H. Caldwell, sent them an excited cable.

But science had to wait until the 20th century for knowledge of the duckbill's intimate life. It is an aquatic animal, living in Australian fresh water at a wide variety of temperatures—from tropical streams at sea level to cold lakes at an elevation of a mile.

The duckbill is well adapted to its aquatic life, with its dense fur, its flat tail and its webbed feet. Its bill has little in common with the duck's. The nostrils are not in the same place and the structure is different, rubbery rather than horny. But the bill serves the same function as the duck's bill, so it has been shaped similarly by the pressures of natural selection.

The water in which the duckbill lives is invariably muddy at the bottom and it is in this mud that the duckbill roots for its food supply. The bill, ridged with horny plates, is used as a sieve, dredging about sensitively in the mud, filtering out the shrimps, earthworms, tadpoles and other small creatures that serve it as food.

When the female is ready to produce young, she builds a special burrow, which she lines with grass and carefully plugs. She then lays two eggs, each about three-quarters of an inch in diameter and surrounded by a translucent, horny shell.

Placing them between her tail and abdomen, the mother platypus curls up about them. After two weeks the young hatch out. Having teeth and very short bills, the newborn duckbills are much less birdlike than the adults. They feed on milk. The mother has no nipples, but milk oozes out of pore openings in the abdomen and the young lick up the milk they need. As they grow, their bills become larger and their teeth fall out.

Yet despite everything zoologists learned about the duckbill, they never seemed entirely certain how to classify it. They finally made hair and milk the deciding factors. In all the world, only mammals have true hair and only mammals produce true milk. The duckbill and spiny anteater have hair and produce milk, so they have been classified, compromisingly, as mammals.

Even so, they're very special mammals. All the mammals are divided into two subclasses. In one (Prototheria or "first-beasts") are the duckbill and five species of the spiny anteater. In the other (Theria or just "beasts") are all the other 4,231 known living species of mammals.

But all this is the result of judging only living species of mammals. Suppose we could study extinct species as well. Would that help us decide on the place of the platypus?

Fossil remnants of mammals and reptiles of the far past are almost entirely of bones and teeth. Bones and teeth give us interesting information but they can't tell us everything. For instance, is there any way of telling, from bones and teeth alone, whether an extinct creature is a reptile or mammal?

Well, all living reptiles have their legs splayed out so that the upper part above the knee is horizontal (assuming they have legs at all). All mammals, on the other hand, have legs that are vertical all the way down. Again, reptiles have teeth that all look more or less alike, while mammals have teeth of different shapes, with sharp incisors in front, flat molars in back, and conical incisors and premolars in between.

There are certain extinct creatures called "therapsids" (order Therapsida) which have vertical leg bones and differentiated teeth just as mammals do—and yet they are considered reptiles. Why? Because there is another bony difference to be considered.

In living mammals, the lower jaw contains a single bone; in reptiles, a number of bones. Because the therapsid lower jaw is made up of seven bones, the therapsid is classified as a reptile. And yet in the therapsid lower jaw, the one bone making up the central portion of the lower jaw is by far the largest. The other six bones, three on each side, are crowded into the rear angle of the jaw.

There seems no question, then, that if the therapsids are reptiles they are nevertheless well along the pathway toward mammals.

But how far along are they? For instance, did they have hair? It might seem that it would be impossible to tell whether an extinct animal had hair or not just from the bones, but let's see . . .

Hair is an insulating device. Reptiles keep their body temperature at about that of the outside environment. They don't have to be concerned over loss of heat, and hair would be useless.

Mammals, however, maintain their internal temperature at nearly 100° Fahrenheit regardless of the outside temperature; they are "warm-blooded." This gives them the great advantage of remaining agile and active in cold weather when the chilled reptile is sluggish. But then the mammal must prevent heat loss by means of a hairy covering. (Birds, also warm-blooded, use feathers as an insulating device.)

With that in mind, let's consider the bones. In reptiles, the nostrils open into the mouth just behind the teeth. This means that reptiles can only breathe with their mouths closed and empty. When they are biting or chewing, breathing must stop. This doesn't bother a reptile much for it can suspend its need for oxygen for considerable periods.

Mammals, however, must use oxygen in their tissues constantly, in order to keep the chemical reactions going that serve to keep their body temperature high. The oxygen supply must not be cut off for more than very short intervals. Consequently mammals have developed a bony palate, a roof to the mouth. When they breathe, air is led above the mouth to the throat. This means they can continue breathing while they bite and chew. It is only when they are actually in the act of swallowing that the breath is cut off and this is only a matter of a few seconds at a time.

The later therapsid species had, as it happened, a palate. If they had a palate, it seems a fair deduction that they needed an uninterrupted supply of oxygen and that makes it look as though they were warm-blooded. And if they were warm-blooded, then very likely they had hair too.

The conclusion, drawn from the bones alone, would seem to be that some of the later therapsids had hair even though judging by their jawbones, they were still reptiles.

The thought of hairy reptiles is astonishing. But that is only because an accident of evolution seems to have wiped out the intermediate forms. The only therapsids alive seem to be those that have developed *all* the mammalian characteristics, so that we call them mammals. The only reptiles alive are those that developed *none* of the mammalian characteristics. Those therapsids that developed some but not others seem to be extinct.

Only the duckbill and the spiny anteater remain near the borderline. They have developed the hair and the milk and the single-boned lower jaw and the four-chambered heart, but not the nipples nor the ability to bear live young.

If we had a complete record of the therapsids, flesh and blood as well as teeth and bone, we might decide that the duckbill was on the therapsid side of the line and not on the mammalian side . . . or are there other pieces of evidence to be considered?

An American zoologist, Giles T. MacIntyre of Queens College, has taken up the matter of the trigeminal nerve, which leads from the jaw muscles to the brain.

In all reptiles, without exception, the trigeminal nerve passes through the skull at a point that lies between two of the bones making up the skull. In all mammals that bring forth living young, without exception, the nerve actually passes *through* a particular skull bone.

Suppose we ignore all the matter of hair and milk and eggs, and just consider the trigeminal nerve. In the duckbill, does the nerve pass through a bone, or between two bones? It has seemed in the past that the nerve passed through a bone and that put the duckbill on the mammalian side of the dividing line.

Not so, says MacIntyre. The study of the trigeminal nerve was made in adult duckbills, where the skull bones are fused together and the boundaries are hard to make out. In young duckbills the skull bones are more clearly separated and in them it can be seen, MacIntyre says, that the trigeminal nerve goes between two bones.

In that case, there is a new respect in which the duckbill falls on the reptilian side of the line and MacIntyre thinks it ought not to be considered a mammal, but as a surviving species of the otherwise-long-extinct therapsid line.

And so, 170 years after zoologists began to puzzle out the queer mixture of characteristics that make up the duckbill platypus—there is still argument as to what to call it.

Is the duckbill platypus a mammal? A reptile? A therapsid? Or just a duckbill platypus?

Condors

by William W. Fox

The steep cliffs and the rocky brush land stretching for rugged miles behind the Santa Barbara-Ventura coastal strip are—as the wags put it—strictly for the birds.

By federal decree some 86 square miles of the spectacular vertical country is set aside for the exclusive use—in perpetuity—of the California condors. The preserves are known as the Sespe and Sisquoc Condor Sanctuaries.

The condors themselves appropriated these two bits of real estate many bird generations past as their most favored homelands.

Some 83 square miles in the Sespe Sanctuary constitute the condor "bedroom," sheltering about 25 nesting sites, almost three quarters of the total number known anywhere. The California condor nesting taste runs predominantly to the cave-in-a-cliff motif, although one instance is on record where condors occupied a hollow in a tall tree.

The three-square-mile Sisquoc Sanctuary contains a pond, clear and shallow, in an obscure high valley that opens onto an ideal launching drop where a small waterfall cascades over the lip of a cliff. Here the big birds bathe, preen and soar.

Actually, *Gymnogyps californianus* is an endangered species, the total population having shrunk to about 60 individuals. The main reason is crowding. The two sanctuaries, then, would seem to form an ideal nucleus for the last-stand habitat of the shy birds.

Visually, the California condor and its Andean counterpart are similar. However, they are not only different species but representatives of different genera, the South American bird being *Vultur gryphus*.

Gymnogyps californianus is one of the world's rarest birds as well as being North America's biggest free-flying land bird. An adult stands some 2½ feet tall. The largest actually measured was nine feet seven inches from wing tip to wing tip.

The condor's head is orange-yellow and essentially bald, which accounts for the "Gymno" in the genus. Outsize feet add to the general unloveliness of both male and female, which humans cannot tell apart visually.

Viewed from below in flight, the adult birds are unmistakable because of the white triangle that occupies about half the wing, behind a dark margin on the leading edge.

In 1953 the National Audubon Society published Carl Koford's condor study which included a population estimate in the same general range as today's counts. There is no strong indication of recent increase, nor any prospect of one. In fact, biologists believe that a 30 percent decrease has taken place since 1930.

A reasonable assumption is that the population was once much larger because it occupied a broader range. In the early 1800's the birds were reported thriving from British Columbia to Mexico and east into Wyoming. Lewis and Clark saw them along the lower Columbia River. The fossil record extends the range during the Pleistocene to Texas and even to Florida.

Eating is not a regularly scheduled event in a condor's life. They are, in fact, thought able to survive a month without food if they must.

Some observers believe condors gorge themselves so heavily when they have the opportunity that they cannot get off the ground. Others think the post-prandial grounding comes from eating food they cannot digest.

A condor meal is seldom neat and orderly. The main course is likely to be sheep, a cow or a deer that previously has died. Ground squirrels and other small animals are eaten from time to time.

A feast on a cow may be more like a brawl than a banquet. But the vigorous tugging and hauling must be blamed on bad tools. The condor beak is not a sharp instrument for penetrating a cow's hide, so the birds must draw chunks of meat out through the natural openings, or through openings they are able to make via soft spots.

Normally, golden eagles will be the first to feed on dead animals. Other birds await their turns at the risk of injury if they don't. When the eagles have left, condors lumber in; then come the buzzards, the ravens and the ants, in that order.

Condors have a peck order of their own that becomes apparent in their phase of the feast. Mature, strong adults take the first turn at the food until they are satisfied. They are followed by the young and weak adults. The location close to the Pacific Ocean is important because of the prevalence of onshore breezes. These strike the wedge of high coastal hills and rise, forming the up-drafts that carry the birds to favorite altitudes.

Once aloft, the condor sets its heading and uses the up-drafts for his long, majestic soaring flight in search of food. He may soar 10 miles on one bearing without appearing to move a muscle.

Descending from the sky is no problem unless the bird urgently wants to return to earth to feast on a carcass he has spotted. He merely lowers both wings and descends in a fast, steep dive. Other birds of prey, the eagle for example, use an altogether different maneuver that Koford has termed the "double-dip." In executing it the bird tips up almost to a stall, then goes into a steep dive, leveling off for another stall and dip. Why the stall? The bird is not built to withstand the "negative G's" that would be imposed upon it in a straight nose dive.

Despite charges of mayhem and theft leveled by sheep men, particularly, the California condor is as harmless as it is charmless. The fact is that the bird is ill equipped to commit vicious acts. Its beak is not formidable and its big feet are adapted only for walking. The toes are physically incapable of grasping like eagle talons. They even lack the capacity to roost on branches. As a result, the condor roost is usually on a surface broad enough to give stability without having to latch on. Condors do have, however, an innate sense of balance which enables them to perch on the most slender limbs.

Much of a condor's time is spent preening, sunning and splashing water over itself in the shallow bathing pools. It suns itself frequently, standing upright with its wings outstretched to get the most out of the warming rays.

Apart from the enforced inactivity of egg sitting, condors often appear simply to be living out their time. They may sit on one perch or move from one to another without going aloft for a day or more at a time.

Only a mated pair ties itself down to a nesting site. They may forage 40 or 50 miles away but they always come back to their chick. Others seem to have base areas but may roost overnight wherever the urge overtakes them, provided the conditions are right. Even a one-time roost must promise freedom from human disturbance and must have an easy escape route into the air by dependable up-drafts or by drop-launching from a high spot.

The spot selected by a pair for raising a family may not be a cave. It can be a cleft in the rocks or a protected space behind a boulder. Ordinarily it is high above the valley floor and is always remote from man. It must have a flat floor because the parents will erect no barrier to protect the egg from rolling out. A mated pair may produce a single egg every second year. Only eight condor chicks are known to have been born during a recent six-year period when condor biologists were giving the subject close attention.

How slender the thread!

Both parents serve during the 50-day period until the egg hatches. In its first weeks as a chick, the young condor's appetite is tremendous, averaging a daily consumption rate of 2 lbs. of meat per bird. At six weeks the bird resembles a down football with a scrawny head and big feet.

The life of *Gymnogyps californianus* is filled with hazards. A partial list must contain the following: 1. Encroachment by man in developing the land; 2. Fire prevention and suppression activities; 3. Uncertainty of food supply; 4. Poisoning; 5. Snooping and photography; 6. Aircraft over-flights; 7. Shooting.

Emergence of the militant environmentalist may work to the condor's advantage. One case is recorded where a road was proposed near the Sisquoc Sanctuary. Bird lovers protested and the road was not built.

Sisquoc and Sespe Creeks are well inside Los Padres National Forest so their continuance is assured for a while. But as a special interest group the condors and their partisans have been only moderately impressive.

Much of the 15-20,000-square-mile range that represents the whole habitat is privately owned. Management decisions by individuals or corporate owners may be for the birds or against them. Some will be made by men disinterested in *Gymnogyps*.

Wildfire in 1972 burned off valuable watershed cover and threatened to engulf the Sespe Sanctuary. Some of this costly tragedy is chargeable against condor protection. Was the benefit worth the cost?

Heyday of the feathered scavengers must have been in the days when the livestock business was conducted for hides and tallow. The rest left to the condors. Time has altered the livestock business. These days, nothing is discarded.

Food supply is not critical for the condors, according to biologists. They have started slipping highway-killed deer into the nesting areas where the parents can get it easily. Results should show in a few years.

Poison is a diminishing threat. Intensive livestock-oriented programs of poisoning have been directed against rodents and predators. The poison entered the food chain one or two links from the condor: one, if the bird fed on the poisoned squirrel, two, if it fed on the coyote that ate the squirrel. Condor mortality rate from this cause has dropped.

Egg collecting and condor capture are prevented by law, as they are major impediments to population growth. Photography and snooping within the sanctuaries are minimized by custodial vigilance, because these activities keep birds away from nesting duties. The Forest Service encourages viewing from a distance when the birds are at their best, high above condor country. Two observation points can be reached by automobile if one is willing to drive several miles from the freeways.

Public information is generally credited with reducing bird slaughter by firearms. There is no open season on condors and that fact is posted and publicized. Little known violence has occurred in the past few years. The Forest Service, the Bureau of Sport Fishery and Wildlife and the National Audubon Society have stationed specialists in Ojai or Ventura, largely to keep the public informed.

These men believe firmly that only an informed public can strengthen the slender thread that preserves this soaring giant.

Leopard Cats

by Caroline T. Sullivan

A new type of Asiatic beauty is invading the Western world. Coming on "little cat feet", assisted by jet flights, leopard-like cats the size of domesticated cats have been creeping into the pet shops of the United States.

With their striped faces, spotted bodies and striped tails, the cats look like diminutive leopards. Like all wild cats, they have a pouch—not a marsupial type pouch—but a pouch of excess fat on their undersides.

They are imported from India, Indochina and Indonesia. Similar, although not necessarily related species, are found throughout almost all tropical countries. The species reaching the United States is commonly called the "leopard cat" (Felis prolinarius bengalensis). It was named after the first one recorded in the West was found in the area of the Bay of Bengal in 1792.

Little has been written about the leopard cats although they are very common in the Orient. Their small hides seem to have precluded an interest equal to that in the larger wild cats. They are rarely found in zoos because their nocturnal habits and extreme timidity prevent them from being the exciting display animals that their beauty warrants. Nocturnal, and either arboreal or living in caves in their native lands, these feral cats have difficulty adjusting to captivity.

Despite this, enough have been imported to warrant their

registration by the Cat Fanciers' Association Inc. (a first for an exotic breed) and the formation of a Leopard Cat Information Center in Los Angeles. Mrs. Virginia English, the owner of several leopard cats, organized the information center in an attempt to discover more about the animals and to aid other owners of leopard cats.

Semiaquatic in their native haunts, the cats have the distinct advantage of being easily trained to excrete in a water-filled pan. Some can be trained to use the commode.

Frequently this becomes the only advantage. Purchased at scalper's rates from pet shops with the fallacious belief that these wild animals can be easily tamed and will materialize into "normal" housecats of exotic beauty, the cats are taken into homes ill-equipped to give them the environment they require. While not normally vicious, their timidity makes them ferocious when cornered or pursued. The distinctive cry of the leopard cat can be quite frightening to someone accustomed to the gentle "mew" of the domesticated cat.

Brutal treatment is sometimes given to the cats by disappointed owners. More frequently people ignorant of the social nature, special feeding requirements and toilet habits of the species inadvertently cause much suffering to the animals and fail to receive the anticipated pleasure from their ownership.

If properly handled by people who are willing to adjust their home life to the cats, Mrs. English assured me that they can make interesting and beautiful pets and occasionally quite affectionate ones. She cautioned, however, that this is very rare with animals born in the wild or even with domestic-born leopard cats if they are not hand raised. She isolates her kittens from the mothers by the time they are 48 hours old and bottle-feeds them. The resultant kittens seem to be much gentler than either the foreign or domestic born which are raised by their mothers.

When acquainted over a period of time the cats are easily bred in captivity, although the litters are small—two being the usual number. They welcome new leopard cats, even males of eligible age, into their groups. However, they are standoffish, although noncombatant, with cats of other species. Occasional mating with domestic cats will occur and because of the nasty disposition of the resulting offspring, the Leopard Cat Information Center encourages leopard cat owners to have any other type of cat they may own either spayed or castrated to prevent this unfortunate mixing.

Besides deliberate mistreatment, the leopard cats also suffer from malnutrition with frightening frequency. Raw beef chunks, raw chicken necks, and cottage cheese with an egg yolk added occasionally, were recommended by the information center. Additional supplements of vitamins and calcium throughout the cat's lifespan are regarded by the center as imperative.

Because of the leopard cats' foreign origins, they also need to be protected from diseases to which our native cats are resistant. Aside from the usual inoculations, the center strongly recommends that leopard cats be kept in their homes at all times to prevent exposure to disease and also so they will not wander away or be accidentally injured.

Not only does the Leopard Cat Information Center provide kits with health and inoculation certificates but they provide an unusual service for leopard cat owners in the Los Angeles area—a blood donor service. Transfusions are available in emergencies with the blood from healthy leopard cats.

Although modern science has fortunately provided blood transfusions, there is no available technique of artificial insemination. As with other cats, ovulation does not occur until stimulation by the "barbs" on the male cat's penis. At present, there is no way of duplicating this process. This is a blessing for the species. If artificial insemination were practicable, it undoubtedly would be used by some breeders knowing that hides from the little leopard cats would have a ready market in the form of fur coats and accessories.

There is little information available about the habits of the leopard cats in the wild—indeed there seems to be little upon which various experts can agree. The careful breeding and feeding files maintained by the Leopard Cat Information Center may help fill this gap. Even the anecdotal information, contributed by various owners, is being carefully processed to pick out verifiable characteristics that may be common to the species. Insight into the jungle life of the cats may be gained in this manner, although environmental influences will necessarily change many of their observed characteristics.

Mrs. English has attempted lobbying to prevent the further importation of these hapless creatures. However, educating the public to the disadvantages of having a wild cat in their homes seems to be a speedier method of preventing further death and injury than pushing legislation through an already overloaded maze of federal, state and city bodies. Leopard cats that have survived the trauma of capture and shipping have the potential of preventing similar suffering by other jungle cats; properly utilized for breeding, they can supply the demand for domestic born, hand-reared cats. People too impatient to cope with being put on a waiting list for domestic-born-and-raised leopard cats can have their pick of ordinary kittens, available through humane societies and advertised in newspapers. By acquiring these, a person is preventing rather than encouraging suffering of leopard cats.

Seamamm's Amazing Pinnipeds

by Ellen R. Hartley

On a sunny but stormy day in January, 1972, the Explorer II stood out from its home port of Key Largo, Florida and headed northeast toward Freeport on Grand Bahama Island.

The 64-foot yacht-oceanographic research vessel of the Sea Mammal Motivational Institute (Seamamm), a land- and sea-based research organization studying sea mammals, was manned by a young couple, Nina and Robert L. Horstman, founders and directors of Seamamm.

The Horstmans and their crew, 17-year-old Susie Redmond, a student volunteer, 12-year-old Jeffrey Horstman, his sister Pami, 11, and kid brother Andy, 4, were on their way to Freeport to recover Tinkerbell, their eight-year-old California sea lion with a penchant for making human friends. The research animal had been missing until the U.S. Coast Guard reported seeing it in Freeport harbor. A staff member made a positive identification, and the Explorer was on its way to bring "Tink" home.

When Tinkerbell saw the vessel she flipped right aboard, using a ramp installed for the animals, and settled on the carpeted afterdeck, happy to be home.

The return trip was a nightmare. The Horstmans battled 30-foot waves while their young crew writhed from seasickness. At last, weary from the nearly sleepless crossings, Bob and Nina brought the Explorer into Seamamm's land base, the Ocean Reef Club on Key Largo.

Here, after all the trouble and expense, Tinkerbell was turned loose again. Experiments under restricted freedom are, perhaps, the most amazing aspect of Seamamm, one of America's newest and most unusual animal behavior laboratories.

It began in 1965 when the Horstmans occupied a beach house in Longport, New Jersey. Nina thought a seal, permitted to come and go as it pleased, might make a logical addition to their menagerie of cats and dogs.

It took persistence to convince Bob, but eventually he agreed to join Nina in studying animal behavior, with special attention to seals. When his job permitted, the couple travelled around the country to discuss pinnipeds—flippered sea mammals such as seals, sea lions and walruses—with people who kept them.

Bob Horstman says, "We were lucky to bump into a professional tree surgeon, Harry A. Goodridge, who had captured a harbor seal *(Phoca vitulina)* and named it Andre. Eventually Harry became an authority on seals in their wild state."

Storms and natural enemies, including man, often separate seal pups from their mothers, and some may simply be deserted. In May 1967, Goodridge helped the Horstmans find a harbor seal pup on the coast of Maine, a 25-pound male a few days old. They named him Shag.

Without books on raising seal pups, bringing up Shag became a full-time job. The few "experts" who had succeeded in keeping baby seals alive had used trial-and-error methods. The Horstmans studied all of them. Shag developed a habit of licking a mixture of cream and egg yolks from their ankles, but he lost weight steadily. Unless he coud be made to eat fish, he would die.

In desperation, Nina rolled Shag on his back one day as if to tickle his stomach, a caress he loved. Instead, she pushed a tiny slice of mackrel down his throat. Surprised, he swallowed—and demanded more.

Teaching Shag to live in freedom was another problem. He

refused to stay in the ocean without his "family." One August evening, however, he hesitated to leave the water and the Horstmans rushed home without him. Next morning, trails indicated Shag had come home during the night—but had then returned to the ocean. From then on he came and went as he pleased. Shag remained with the Horstmans for three years before dying of intestinal infections.

Shag taught the Horstmans that seals, given complete freedom, will eventually return "home."

The Horstmans decided to establish a sea mammal motivational institute. Reputable scientists would be asked to provide guidance by forming the institution's advisory board.

In 1969, Bob and Nina brought Explorer II, specially designed as their sea base, to Key Colony, Florida. The vessel has a range of 4,400 miles and is automated so that one individual can operate it. Aware that planned scientific activities required a land base, they searched for suitable permanent quarters. Seamamm moved into its new home, The Ocean Reef Club on Key Largo, in January 1972. Officially formed in February 1970, Seamamm was granted tax-exempt status as a public foundation funded by private donations.

Although funds are important, one of the organization's first donations was equally welcome. It was Tinkerbell, who had been captured when about two years old, and had spent four years in captivity. Tinkerbell arrived distrustful of humans. Friendly advances by the staff were answered with threatening bellows. Finally she ambled over to resting nine-year-old Pami Horstman, climbed onto the air mattress next to the girl, and made friendly overtures. Two months later "Tink" was playing with the Seamamm staff.

Donated to Seamamm in September, Tinkerbell was permitted full freedom in January. A mature animal, capable of supporting herself, she sniffed the air and headed straight for the water. Soon she moved out to sea. After a week-and-a-half, when Seamamm had almost given her up, a sighting was reported by a radio station in the Florida Keys. A few days later,

Andean Condor

Whooping Cranes in pursuit of territorial invaders

Whooping Crane chick

Wolverine preparing to attack

Prowling Wolverine

Giraffe and calf

Uganda Giraffe and Zebras in natural habitat

Flat-tail Tortoise

Eohippus—the first creature in history that could properly be called a horse.

Mountain Gorilla

Lowland Gorilla and infant

Lioness and cubs

Tiger-sized yawn

Care and feeding of domestic Marmosets (top and bottom)
Cotton-top Marmoset (right)

Leopard Cat (kitten)

Tinkerbell was seen near Seamamm facilities. Staff members drove along the beaches. "Tink," floating offshore, recognized her friends and without hesitation, hoisted herself out of the water at a boat ramp and got into the waiting van.

Tinkerbell now stays with Seamamm by choice, leaving whenever she desires. Some of her independent trips have taken her from the Fort Jefferson National Monument in the Dry Tortugas, about 60 miles southwest of Key West, to Boca Grande, 40 miles south of Sarasota. She has journeyed to the area of Washington, North Carolina, where people named her "Samantha," and radio stations reported on her whereabouts during the regular daily weather forecasts. On another trip, 167 miles north of "home" she surprised a vacationing family by plopping down beside their swimming pool.

When we visited Seamamm in February we met Tinkerbell, Vicki, a 2½-year-old sea lion, and Rocky, a harbor seal now about two. Vicki and Rocky joined Seamamm as pups. About 10 other animals will be added to the collection during the next few months.

Vicki also made national headlines when she left Seamamm recently. Bob says, "We were working with the animals from the Explorer out on the reef. Vicki swam to shore and joined a family swimming and camping there, to the delight of their young daughter who loved her unexpected swimming companion.

"They were responsible people and took good care of her. She slept with them in their tent. When they departed they took Vicki along and left her at the Miami Seaquarium. Had Vicki been left to her own devices she would have returned to the land base or come back to the Explorer."

So far, Seamamm activities have hinted that seals may provide valid behavioral and physiological information that can help man exist in an alien environment—the ocean. As marine mammals, they have made all necessary adaptations for pelagic existence.

They communicate, and they form an affinity for humans. Rocky, for example, uses a toy rubber ring to tow swimmers toward land. It seems natural fun for him to tow anything tied to his ring—lines, gear, people, even a small boat. Since other seals share this trait, they might be useful for beach guard duty. Seamamm animals are trained to take a flotation device or a line to swimmers in trouble.

Their training provides other examples of human and pinniped collaboration. We watched the animals "work" over the reef from aboard the Explorer. A staff member stationed himself on the bottom of the ocean while another told Tinkerbell and Vicki to take tools and lines and other equipment to him. To the animals this was a game. They responded eagerly, and returned other objects to the humans on the boat as directed by the man below. This is not a scientific breakthrough; porpoises (dolphins) have done it and the U.S. Navy has seals working in a free-release pattern.

Seamamm's intelligent pinnipeds have figured out the basic principle of banking. Rewarded with a plastic disk for each accomplished task, they will either exchange right away for some fish or collect disks and exchange them later on—sometimes days later.

The writer watched Tinkerbell try to "rob" the plastic disk bank by attempting to open the box. When that proved futile and the refrigerator in which fish is stored also withstood her probing nose, "Tink" searched for plastic disks "banked" by Vicki. She stole one, got her fish—then went back to earn her own disks.

Later this year, in a new experiment, plastic disks of different shapes will be introduced to the animals. The Seamamm staff calls them "designators." Each shape will have a different meaning. One may indicate "I want petting," another "swim with me," still another, "let's play," etc. The animals will have free access to the designators.

Scientists on Seamamm's advisory board are planning their own experiments. Dr. John P. Kerr, associate professor of biology and marine sciences at the University of West Florida in

Pensacola, hopes to investigate how sea mammals use their sense of smell.

During our first meeting with Vicki, she rushed up to us, pushed her nose and whiskers to our mouth and sniffed. Then she buried her nose in our hair. Apparently accepting us, Vicki demanded to be petted—then jumped back into the water.

During our second meeting, aboard the Explorer, Vicki again sniffed our mouth and hair and obviously recognized us. The animal had no difficulty communicating its requests: "jump overboard and play with me; pet me; if you don't want to swim, throw your cup of tea overboard. I'll retrieve it."

Sniffing breath may look like "kissing," but it is the pinniped method of identification. Mother sea lions may recognize their pups by smell.

Dr. Ralph Stolz, a member of the Aerospace Medical Association and a member of the Undersea Medical Society, will head a team to compare body functions in Seamamm divers with those of their diving sea mammal companions. Men and pinnipeds will swallow tiny transmitters to feed back biological information through use of bio-telemetry.

Seamamm divers and animals have already worked at a depth of 75 feet. However, sea lions reportedly have gone to a depth of 750 feet. They can do this because they can slow their heartbeat. They conserve oxygen through constriction of blood vessels and can limit circulation, giving heart and brain most of the available oxygen. Scientists who had considered these actions involuntary in humans are now realizing that man, too, can learn to slow his heartbeat.

This is only a sample of planned experiments, but a promising beginning for the only private institution committed to free-release work with sea mammals. Seamamm receives no public funding and is entirely dependent on voluntary contributions.

Nina Horstman says, "We literally live, work, eat and sleep with our animals. Their affinity with humans is the only tool we use to keep our animals with us. Because they have access to the open ocean, we have a 24-hour responsibility to keep them intellectually satisfied." She believes the daily logs kept on the animals provide the most complete information on pinnipeds under free-release conditions.

Bob Horstman says Seamamm will establish herds of seals, sea lions and dolphins which will live unrestricted around the land and sea base.

Bob adds, "The Sea Mammal Motivational Institute is here. What we need now are scientists in specific disciplines, particularly behavioralists, who will join us and design their own projects, using our facilities.

"We want to find universities and foundations willing to fund research projects. In turn, we will supply students and scientists with an environment of free-release sea mammals for their study."

Seamamm staff members believe that when man works with an animal he is responsible for its physical and psychological well-being. When a report on activities appears, the Horstmans are swamped by offers from people like Kathi Ogden, a 19-year-old sophomore from the University of South Carolina. Kathi earned 13 credit hours for the six months she worked at Seamamm last year.

Bob Horstman says, "We want the help, but can't strain our financial resources. Universities will have to provide funding for research and the instrumentation graduate studies may require.

"We hope someday, when Seamamm has grown, to have a special department of education to absorb these expenses. Right now, we must keep a rather limited staff.

"Students will get top priority if they come to me and say, 'Such and such a university will give me credit hours and I will have to report on my work, or in the case of a postgraduate student, 'publish my report'."

The Horstmans have been so delighted by Seamamm's success in returning Tinkerbell to her natural element that a special fund has been established to return captive sea mammals to the ocean. For example, there's a dolphin kept in a Maryland pet shop in a small collapsible pool. Bob says, "It's obscene. Dolphins are animals that must communicate with their own kind. To keep them in solitary captivity is a crime.

"We are committed to take animals out of these restrictive environments and reintroduce them to the ocean when fund contributions reach $15,000."

Bob, as an advisor to the U.S. State Department, recently flew to London where an international treaty on the harvesting of seals in Antarctica has been worked out.

He says, "Many of us felt Antarctica should be a sanctuary, that there should be no harvesting there. However, some nations insist on sealing, so without some kind of international treaty the animals would have no protection."

Domestically, several bills to preserve sea mammals are pending in Congress. These would regulate the taking, transportation and keeping of sea mammals.

In September 1971, Bob testified before the House Subcommittee on Fisheries and Wildlife Conservation. He would like to see a moratorium of at least five years (ideally 10 years) on the taking of all sea mammals until we know how many are left.

Bob points out that *effective* marine mammal legislation must be well conceived and written. Nobody who has seen Seamamm's amazing pinnipeds would want them and their cousins to follow the dodo into extinction.

The Meanest Animal in the World

by Ronald Rathert

Two weary trappers plodded toward their tiny cabin, their clumsy snowshoes plunging awkwardly through the ice-glazed snow with every step. Suddenly the older man in the lead stopped and, wrinkling his nose in digust, whispered to his companion: "Skunk bear."

Hurriedly tugging at his friend's white parka while gesturing for silence, he swung his Winchester 73 from his shoulder and began creeping down the slippery trail.

Two minutes later they were surveying the shambles of their simple cabin. Log walls two feet thick had been gnawed through, pots and pans had been strewn about, bacon, oatmeal and canned goods had been eaten or torn apart and soiled with urine.

This scene has probably been repeated thousands of times throughout the world. The tired outdoorsmen had been unwilling hosts to a fabled animal: the legendary wolverine.

Today, the wolverine is dying. Individually he remains his fearless self, but like the passenger pigeon and the whooping crane, this scrappy little swashbuckler has been outmoded by civilization and is in danger of becoming extinct. Michigan, nicknamed the "Wolverine State," has been abandoned by the species.

With the humpbacked posture of a grizzly bear, the tremendous, muscle-laced forelegs of a panther and the face of a balding demon, the wolverine sports a body about three feet long and a foot high at the shoulder and, when full-grown, he weighs about 50 pounds.

This fiery, unbelievably powerful little beast goes by a variety of names, most of them derogatory. Americans sometimes call him the "skunk bear" because of his resemblance to a combination thereof; in Europe he's known as the "glutton," while French-Canadians have named him "carcajou" and Eskimos revere him as the "Quiqui-hatch" (invulnerable beast).

About 25,000 years ago the Asiatic wolverine made his way to the North American continent across the land bridge now flooded by the waters of the Bering Strait. Somehow he also reached central and east Africa, evolving as a smaller cousin, the *ratel,* a relentless enemy of native herdsmen.

Most adult wolverines, born silvery white, are dark brown or black, with a pale yellow streak of fur running from shoulder to rump. Their fur is long and shaggy, covering even the soles of their paws.

The range is dwindling; once roaming most of North America and a good portion of the Old World, the wolverine is rapidly fading into the frozen tundra and evergreen forests of northern Canada, Scandinavia and Siberia.

No animal hunts the wolverine; he fears no creature, including man. The largest of the weasel family (mustelidae), the "skunk bear" will eat almost anything: berries, rotting salmon, mice, mountain sheep, deer and chipmunks. He's been seen driving mountain lions and grizzly bears away from their kills and gorging himself on their meals. He enjoys the infuriating habit of following trap lines, eating the bait and captured animals, spoiling what he can't eat and destroying the traps.

The wolverine's remarkable courage, strength and cunning give rise to a romantic aura:

> "And by the well-worn path the Carcajou
> Drops from his hidden perch upon the unwary prey."
>
> —Duncan Anderson

The British zoologist Thomas Pennant wrote shortly after the American Revolution that wolverines "drop moss from tree limbs to lure deer into ambush."

North American Indians and Eskimos worship the "invulnerable beast," offering sacrifices in his name. However, when they occasionally capture one, they put out his eyes and mutilate his body. Then he's set free, since the demon can no longer cause harm. Their respect for this animal is almost mystical; only one prehistoric carving of the wolverine has been discovered, perhaps because early men often believed that to reproduce the likeness of a god is to invite misfortune.

"Carcajous" are solitary animals, mating briefly in hidden underground dens and producing from one to five wooly cubs in early spring. The young are abandoned in late summer to fend for themselves.

Strangely, for all of his fierceness the wolverine makes an excellent pet when raised from birth. His devotion to his master is as great as that of a German shepherd. Who could wish for a better companion on a dark night?

And the wolverine is peculiarly adapted to the black night. Suffering from snow-blindness, he seldom ventures forth in daylight except when in dire need of a kill. Then the rascal stands upright, shading his eyes with a forepaw while scanning the distance for prey, resembling a diminutive Abominable Snowman.

While wolverines prefer private lives and fight bloody battles with each other upon inadvertent encounters, in desperate circumstances they occasionally team up to drive reindeer into narrow defiles, using every trick of the big game hunter.

Although of minor commercial importance, the wolverine's thick fur is saturated with a unique oil which repels moisture even after tanning. It's valued by trappers and hunters as trimming for parkas, the sleeves of jackets and leather bags, since frost won't collect on its surface. Between 800 and 1,000 wolverine pelts are taken annually in Canada, but both supply and demand are diminishing. Wolverine hides sold for $21 in the 1920's; today the average price is not more than $4.

Peter Krott, Finnish naturalist, writer and animal dealer, discovered that wolverines are in some ways surprisingly delicate animals. Two cubs he bought from a Lapp reindeer herdsman became seriously constipated by the diet of undiluted boiled milk he fed them. He learned later that their tiny bellies and rears must be gently massaged while they suckle, as a substitute for the care of their mother who applies her tongue for the same purpose. This process is virtually essential for the proper functioning of the pups' digestive systems.

Shortly afterward his affectionate, teddy bear-like pets developed chronic diarrhea and died.

Wolverine babies are also subject to *xerophthalmia*, a deficiency condition caused by the lack of vitamin A. Ultimately, it results in total blindness and paralysis of the animal's hind legs.

Yet despite the natural and man-made odds against him, the wolverine remains the "Demon of the North."

He matures more quickly than the other carnivorous animals of the forest and soon becomes almost invulnerable to anything but a bullet. He can even survive poisons such as strychnine. His antiseptic saliva helps his wounds heal quickly.

Within a few months after birth the ferocious little "skunk bear" has learned to swim and has mastered the art of tree climbing. While his hearing and eyesight are comparatively poor for a meat-eating mammal, and though he runs awkwardly and slowly, the wolverine's stamina is amazing.

He stalks and tries to kill every animal who foolishly crosses his path. His impulsive acts of destruction are committed without malice but not necessarily because of hunger. Peter Krott mentions that one of his well-fed pet wolverines managed to force his snout through the bars of a cage housing a heron and ate the bird slowly as he gradually pulled its body from the enclosure—alive.

The exploits of the wolverine are legendary and considerably fictional, but every myth contains a grain of truth. The two men mentioned at the beginning, trapping beavers, minks and otters in the rugged Canadian Rockies during the late 1800s, fought a long-standing feud with one particular "skunk bear," with almost comical overtones.

This animal had broken into their cabin several times, destroyed their traps and eaten or defiled the bait and captured animals. Exasperated, the two trappers placed a steeljawed trap in the fireplace and packed snow banks tightly against the walls of their cabin. Later, while they were checking their damaged trap lines, the wolverine tumbled down the cabin's narrow chimney. He lost half a paw in the jaws of the trap, stole the slab of bacon they had left as bait and smashed his way out of the cabin.

Within a few years his kind will probably be finished forever, but men will always feel a grudging respect for his spirit.

How to Catch a Hippo

by Robert P. Crossley

To catch a hippopotamus, the first thing you need is a road grader—so you can drive out into the river and pacify the beast with a central nervous system depressant called Sernylan. Mixed with another drug called chlorpromazine, the dose is administered with a dart shot from a crossbow.

When you drive the grader into the river, the hippos will raise up out of the water to see what's going on. That's when you choose the one you want, usually a young specimen weighing about 1,000 pounds. You climb up on the cab, cock the crossbow and aim for the animal's neck where the drug will have the most immediate effect. In 10 to 15 minutes this specimen will become groggy, swim slowly in circles or stumble into shallow water.

In order to prevent the animal from faltering and possibly drowning, Sernylan is used instead of quicker-acting "knockout" drugs used on elephants.

Next, you guide the grader over to the passive hippo and throw a lasso around his massive neck. Then you lash him to the side of the grader and haul him to shore. Before crating the hippo, his dart-wound should be disinfected and 12 million i.u. of penicillin should be injected into the animal.

If his grunts bring the mother hippo or the herd bull comes charging after you, the grader's motor should be raced so they can't hear him.

That's the way they do it in Kruger National Park, South Africa, where 2,200 hippo live more or less happily with 7,700 elephants, 16,695 zebra, 2,700 giraffes, 15,300 blue wildebeests, 5,000 kudu, 132,500 impala, 15,800 buffalo, 1,200 lions, 650 leopards, 260 cheetahs, 4,000 warthogs, and other antelope, baboons, monkeys and assorted predators. If these figures seem extraordinarily precise, they are. To manage one of the world's greatest herds of wild animals in the park's 40-by-200-mile expanse of low hills, scrubby woods and grasslands stretching along the Mozambique border from the Limpopo River south to the Crocodile, Dr. U. de V. Pienaar, the park's chief biologist, and the rangers who work with him, have to know exactly how each species is doing. They count them by helicopter.

Because no hunting is permitted, and because diseases have been controlled, and because dams and wells provide plenty of water, the population of most species has literally exploded.

There were no elephants in the area when the original game reserve was established in 1898. Poachers and ivory hunters had wiped them out. By 1905 a herd of 10 had roamed across the border from Mozambique.

When the park was officially established in 1931, there were 135. In 1960 there were 1,000. Last fall the rangers counted 7,700 from the air.

The lions, cheetahs and leopards are supposed to keep down the antelope explosion, but they can't keep up with it. Even though the park is six times as big as Rhode Island, surplus animals have to be culled to keep it from being overgrazed. Big ones like elephants and hippos are captured alive and sold or given to zoos and to other parks.

Drugs to immobilize wild animals for safe handling were risky and unreliable when they were first tried 16 years ago.

The first real breakthrough came in 1959 when Dr. Tony Harthoorn immobilized a herd of white rhinos in Natal and moved them safely to Kruger Park, where the species had become extinct.

Today, after much experimenting, Dr. Pienaar and his associates have determined effective combinations of drugs and safe dosages for virtually every species in the park. The one they use the most is M-99 (Propylorvinol hydrochloride) mixed with acetylpromazine maleate. It's most effective for elephants, but it can't be used on hippos because they'd drown, or on the big cats. Another drug, Fentanyl, is used on giraffes. It calms them without causing them to fall. A groggy giraffe can die of exhaustion trying to get up.

A tranquilized animal can be measured or weighed or moved to better grazing. Veterinarians can treat injuries, examine for diseases, take blood and tissue samples and destroy parasites. By marking immobilized animals, biologists can study migration patterns, reproduction, growth rates and longevity. A small transmitter can be inserted so an animal's movements can be followed by telemetry. When culling is necessary, an unconscious animal can be killed humanely, instead of being shot by a rifle with the chance of being wounded but not killed.

The drugs have to work fast so darted animals won't get lost in the tall brush before the tranquilizers take effect. Drugs like these must have a wide margin of safety, because it's hard to estimate the weight of big individuals to prepare a precise dose. And drugs have to be potent in relatively small amounts so they can be loaded into a dart.

Elephants present the biggest challenge, because they're so big and so fierce, and because they're so prolific, increasing at an unbelievable 11.3 percent a year. They knock trees down to get at the tender top branches and chase smaller animals away from waterholes in dry weather.

Capturing an elephant is a little more complicated than catching a hippo. Instead of a road grader, Dr. Pienaar climbs into a three-place helicopter from which the doors have been removed so he can lean out with his crossbow. From the copter he has two-way radio communication with a ground party in Land Rovers and pickups.

The pilot takes him on a quick reconnaisance flight to locate a herd of elephants. When he finds it, Dr. Pienaar directs the ground party to the area. The helicopter lands nearby so he can confer with the ground party. Then it goes aloft again. Dr. Pienaar picks out the calves he wants to capture or examine, cocks the crossbow, leans out of the copter and deftly puts a dart into the animal's neck.

The helicopter moves higher and away a bit so the herd will calm down. In 10 to 15 minutes the calf will collapse. The herd stands around, trumpeting excitedly and throwing up clouds of dust while the mother tries to lift the calf bodily with her trunk or roll it to and fro with her foreleg to wake it up. If she notices the dart she may pull it out with her trunk and trample it.

Now the helicopter comes back and by buzzing the herd chases away all but the anguished mother. The latter stands over her fallen calf after the rest of the herd has gone, rearing up on her hind legs as if to snatch the helicopter out of the sky. Only repeated passes with the copter drive her away, and even then the copter has to remain on guard as the ground crew moves in with trucks and crates. A ranger ties a nylon rope around the calf's neck and passes the other end of the rope through the open door of the crate and out the back. Another warden gives the calf a shot of an antidote called M-285. In two or three minutes the calf gets up and can be guided gently into the crate. Dr. Pienaar and his aides captured 27 young elephants this way in a period of six weeks, five in a single day. Some were released for being too small or of the wrong sex. They trotted back unharmed to their mothers. Not one of the 27 was killed or permanently injured.

Tourists love to see elephants, but the breeding herds tend to stay away from the roads. Even so, the official park booklet carries this warning:

"BEWARE OF ELEPHANTS. When traveling through ELEPHANT COUNTRY visitors are warned to travel very slowly, especially round corners, and not to try to pass cow elephants with young ones but to turn back as soon as possible, or to remain perfectly quiet until the elephants have gone.

When wishing to photograph bull elephants near the road, drive a little way past the animals before stopping the car and do not stop to take photographs while they are still ahead. Do not stop to photograph a herd of elephant cows and calves but make haste to get away."

Despite all this, Dirk Ackerman told us about a family with two kids who had to walk back to camp when their car broke down. The road was blocked by several cow elephants and their offspring.

"How did you get past?" Dirk asked.

"We threw rocks at them," the father replied.

Dirk broke out in a sweat, just telling about it.

Can the "African Horse" Be Saved?

by George W. Frame

Two zebra stallions fought mightily in the reddish rays of the evening sun. Circling as they neck wrestled, they tried to bite each other with their long incisor teeth. Then one spun around and lashed out with a powerful kick at the other's head, missing by inches. He turned, neck wrestled again, and hung on his opponent in a vain attempt to bring him down. Both were clumsy and panting with exhaustion.

As the fighting became more intense, the challenger grabbed at the harem male's leg, and both stallions fell to their knees. One attempted to stand again, but was continually pushed over by the other. Suddenly both sprang to their feet and circled each other. The challenger then bit the harem stallion's hock, causing him to fall. As he lay still and helpless the challenger chewed away, but no sounds of pain were uttered. Suddenly the challenger released his bite and ran to the harem to claim it for his own. But the harem stallion got up, chased after the challenger and renewed the battle.

I observed this fighting for nearly an hour before the two stallions disappeared into the darkness. During the entire battle, the mares in the harem looked on with intense interest.

A similar instance was described by Dr. Richard D. Estes, who filmed a fight between two common plains (or Burchell's) zebra stallions *(Equus burchelli boehmi)* in Ngorongoro Crater, Tanzania. He watched the fighting for more than two hours, and suspected that they had fought the entire day. Dr. Estes pointed out that the male zebra's eagerness to take on all challengers for his harem is a remarkable example of determination and stamina. If every challenger began fresh, the defender could not have endurance to defeat every rival. However, the rivals also fight with each other, so they often are as tired as the defending stallion.

The plains zebras are one of the most common large mammals in Africa today, although their numbers have declined during the past century. They have a wide range from the Sudan to South West Africa. Several sub-species are recognizable. On the Serengeti Plains of northern Tanzania there are 150,000 plains zebras in an area of about 7,700 square miles. They often mingle with wildebeest, kongoni, topi, black rhinoceros and ostrich. In the annual migration on the Serengeti, they sometimes congregate in herds of several thousands.

During the past few years Dr. and Mrs. Hans Klingel of Braunschweig University in Germany studied the life histories of over 600 individual plains zebras in Ngorongoro Crater. All were identified by photographing the stripe pattern, which is as distinctive for zebras as fingerprints are for humans. Also, 122 were immobilized and marked with ear tags, brands and by cutting the hair of their mane and tail.

The Klingels found that plains zebras formed remarkably stable groups. One type of group is the family, or harem, which consists of one male and one or more females and their young. The other is the bachelor group, made up of from one to several stallions. Family sizes ranged from one to six mares with their offspring, and the largest family totaled 16 animals. Dr. and Mrs. Klingel found that after two years, 95 percent of the females were still living in the same groups.

The female zebra's family instinct is very strong. In fact, mares do not communicate with members of other groups.

When a stallion is displaced, the new stallion takes over the group intact. However, the foals sometimes play with foals of nearby families.

Plains zebra do not mark or defend particular areas, nor do they avoid areas already occupied. There is no social organization above the level of family and bachelor groups. Dr. Klingel says that similar social organization exists in the mountain zebra. *(E. zebra zebra* and *E. z. hartmannae)* in South Africa.

According to Dr. John King, the Grevy zebra *(E. grevyi)*, of northern Kenya, Somalia and Ethiopia is territorial. The largest stallions, up to five feet tall at the shoulder, were found alone on territories of up to one and one-half miles in diameter. Grevy zebra stallions tolerate both males and females within their territories until an estrus mare approaches the boundary; then the stallion fights his rivals for her and tries to drive her toward the center of his territory.

In the 110-square-mile Ngorongoro Crater, where I studied black rhinos and Cape hunting dogs, the 5,000 resident plains zebras gave birth to foals during the entire year. However, 61 percent of the births occur during the main foaling season of January through March. These three months are also the height of the rainy season.

Cohesion of the family group is strong. The stallion protects all his mares which in turn show much devotion to their foals. This was demonstrated last year in Tanzania when a four-month-old zebra was found dead under a tree. Its mother remained with it for six hours and even stood nearby while park personnel examined the young zebra. Death was caused by a bullet wound—a senseless killing by man.

Young stallions leave their families when between one and four-and-one-half years old, before they are mature. They are not chased away by their father. Instead, they leave because of a combination of three things, according to Dr. Klingel: (1) the young stallion's mother has a new foal, (2) there are no other male foals of the same age in the family and therefore no playmates, and (3) other family groups with such playmates are nearby.

Young mares are abducted by either family stallions or bachelor stallions. In this manner, a stallion can increase the number of mares in his family, or a bachelor stallion can begin a new family.

One of the most famous zebras was the now extinct quagga *(E. quagga)*, which usually had white stripes only on the head, neck and forequarters. It was abundant in South Africa until the mid-1800s, when white colonists hunted them. The last wild quagga was exterminated about 1873, and one survived in a zoo until 1883.

The Cape mountain zebra survives primarily in three national parks and one game reserve in southern Africa. The closely related Hartmann's zebra is more numerous and confined mainly to South West Africa.

In 1937, to help preserve the Cape mountain zebra, the government of South Africa proclaimed the Mountain Zebra National Park. Only six animals (five stallions and a mare) were in the park. Although a foal was born, the number declined and by 1950 only two old stallions remained. Then came a turning point. A farmer, J. K. Lombaard, donated his herd of 11 to the park. A fenced passage was built and the small herd was driven safely to its new home. About a year later the first foal was born. Today, the park has 88 mountain zebra.

Lions and hyenas are the major predators of zebras in eastern and southern Africa. In the Serengeti, Cape hunting dogs have been seen chasing zebras, and even a crocodile was seen eating one. According to C. A. Spinage, cheetahs sometimes attack zebras, especially the older males.

The purpose of zebra stripes has been the subject of much debate among zoologists. One possible explanation: most predation occurs during the twilight hours. At a distance, stripes cause the zebra to fade into the background, making it less noticeable. Close up, when the hunt becomes intense, the rapid movement of the stripes helps to blur the zebra's outline and may create a confusing optical illusion.

Poaching is still a problem. Around the Serengeti National Park recently, in just four days of patrolling, a ranger force made 28 arrests, confiscated 264 steel wire snares and destroyed seven large camps. Remains of 162 wildebeest, 33 zebra, 11 topi, four buffalo, and one giraffe were found. Many of the wildebeest had only the tail cut off for ornamental fly wisks to be sold to tourists. Zebras and other animals were killed primarily for meat. During the yearly animal migrations, large gangs of poachers hunt in broad daylight. Every year, in the Serengeti alone, several thousand animals, including zebras, are lost to poachers.

Most poachers travel by car and are armed with rifles and shotguns. But some still use older methods such as pit traps, wire snares and poisoned arrows.

The most modern technique is to poison waterholes. In Lake Manyara National Park, Tanzania, rangers found in 1970 the carcasses of 255 zebras at a poisoned hole, plus hundreds of gazelles, storks, ostriches, hyenas and smaller creatures. The poachers took only the skins and hooves of the zebras to sell as tourist souvenirs.

Danger to parks personnel is increasing, too. Rangers have been killed and are often the targets of spears and poisoned arrows. Organized poaching has become so serious that President Julius Nyerere of Tanzania recently ordered the establishment of a secret service squad to infiltrate the crime network.

But not all the dangers to zebras are so obvious. Most insidious is loss of habitat through cultivation and the interruption of traditional migratory routes. Blayney Percival, who was game warden of Kenya at the turn of the century, wrote that thousands of zebras roamed in many areas. By 1924 he noticed that the movement of zebras had been greatly reduced by wire fences. Today only isolated groups remain.

This was re-emphasized recently by Dr. and Mrs. Klingel: "There is no park or game reserve in Kenya where Grevy's zebra can be found in any numbers throughout the year." During the dry season they must migrate to unprotected areas, and often farmers' fences prevent even this. Another major factor influencing the plains ecosystem is the presence of large herds of domestic livestock which compete with zebras for food.

When wild animal species are allowed to exist without man's interference, they cooperate in a surprisingly efficient manner. In localized areas of the Serengeti, for example, plains zebras feed mostly on the long drier grasses and herbs and then move on. Next come the wildebeest and topi, which feed on the medium height leafy portions of the grasses and herbs left by the zebras. When the wildebeest and topi are finished feeding, they too move on and are followed by Thomson's gazelles, which feed on the remaining short green grass.

As with most species which have declined in numbers in recent years, there is no simple explanation. Hunting and poaching usually are not severe enough to cause a permanent reduction in animal numbers. The big problem facing conservationists is the preservation of intact ecosystems. Areas which contain sufficient food, water and cover during all seasons must be set aside, and these must not be interrupted by fences and cultivation nor intruded upon by domestic livestock and excessive numbers of humans. We must understand and respect the needs of zebras and other wildlife if they are to survive.

Tracking and Killing Maneater Lions

by George W. and Lory Herbison Frame

A lion turned maneater on the dry windy evening of August 23, 1972. He was healthy but hungry, and perhaps braver or more foolish than the rest.

On that eventful night he pushed his way through the dry thornbush fence surrounding a Masai village, while the rest of the pride waited outside. Inside, three eight-year-old boys were sleeping in the open next to their goats. The goats fled in panic, but the children did not react so quickly. The hungry lion killed and ate one of the boys.

This maneating lion made his kill a few miles from our home in northern Tanzania. Several weeks later we learned how the Masai had retaliated. They slaughtered a zebra and put poison in its carcass. Twelve lions ate the meat and died, but whether the maneater was among them we do not know.

Maneating lions are nothing new. Between 1932 and 1947, maneaters in the Njombe area of Tanzania killed at least 1000 to 1500 people. That's about 100 people a year. The lions hunted in three or four small prides over an area of 1500 square miles. During 15 years of maneating, superstition and fear prevented the people from killing even a single lion in retribution. To them, a maneating lion was not merely a dangerous predator, but an animal body inhabited and directed by a living "man-lion" or by the spirit of a deceased person who was returning to life to settle old grievances.

The Njombe maneaters did most of their killing in the evening. Often an entire pride attacked a village. The lions spent most of the night feeding and resting, but at dawn they abandoned the remains and traveled 10 or 20 miles. They separated into pairs and single animals which hunted alone for many nights; then they rejoined for another combined attack. The maneaters never attacked two consecutive nights in the same place.

When George Rushby became game warden of the region terrorized by the Njombe maneaters, he directed his full attention to their elimination. A few fruitless weeks of tracking convinced George that following the lions from the scene of their last kill was hopeless. So, he tried a new approach: improve communications, guess where lions were headed, and try to intercept them. The communication problem was solved by setting up camp along the Great North Road, which passed right through the center of the maneaters' range. George got the latest news about the previous night's kill from the drivers of trucks and buses as they passed his camp. When a trail still seemed fresh, he set out to intercept the killers.

Finally, shortly before noon one day, George and two game scouts came upon the fresh tracks of four lions. They followed the maneaters into thick bush. As the game-scouts looked on with superstitious fear, George managed to kill one female. This marked the beginning of the end for the Njombe maneaters. When the Africans saw that no supernatural harm befell George for killing the lion, they began to lose their fear of the maneaters. In subsequent weeks George and the game scouts shot 17 maneaters, and brought peace to the local population.

No one knows for certain why the Njombe lions became maneaters. Perhaps they had developed a taste for humans by scavenging dead bodies, which Africans traditionally abandoned in the bush. Then one day a hungry lioness may have killed a human because the opportunity was just too easy to avoid; from then on, man became the natural prey. Because the Njombe lions devoured humans over a period of 15 years, it is obvious that lionesses fed their cubs on human flesh, and taught them to hunt for humans. The lions actually developed a preference for humans, and did not bother to kill any of the abundant domestic livestock. Several times Njombe maneaters rushed right through a herd of cattle and killed the herdboy.

Another authority on maneating lions was C. J. P. Ionides. During 23 years in the Tanganyika Game Department he killed nearly 40 maneaters. He hunted one maneater unsuccessfully over an area of 2500 square miles as it ate more than 90 people. The local people were afraid to harm the lion, because they believed it belonged to the witch doctor. In desperation, Ionides finally consulted the witch doctor and paid a fee for a special ceremony to permit killing the lion. A few days later, a young man saw the maneater in the bushes near his village and killed it with his muzzle-loader. Because the witch doctor authorized the lion's death, the villagers were no longer afraid to help themselves.

Both Rushby and Ionides shattered the myth that maneating lions are only the old, sick and starving. Rushby observed that the maneaters he killed were slightly smaller than usual and their fur was glossier and more healthy than that of normal lions. Ionides described one maneater (with 45 human victims to its credit) as a young adult in prime condition, with a fine glossy coat and a good set of teeth. They agreed that lions sometimes kill their first human victim when faced with starvation, but this is most often determined by chance when confronted with an easy kill.

Maneaters are characterized by their boldness. There are numerous incidents of lions breaking down doors, or tearing through mud walls and thatched roofs of huts to get to the people inside.

The most publicized incident of maneating lions occurred in

Kenya at the beginning of this century, when a number of maneaters delayed construction of the railroad. Two notorious maneless males were responsible for at least 58 killings. Other lions killed and harassed the laborers for years.

One determined lion tried to tear through the corrugated iron roofing of the railway stations, but succeeded only in cutting his feet on the sharp metal. Later, the same maneater ate the driver of one of the railway engines. Another engineer, hoping to shoot the lion, spent all of the next night watching for the lion from the safety of an iron water tank. The wary lion discovered the ambush, overturned the sturdy tank, and tried to attack the engineer through the entrance hole. But it was driven away when the engineer fired his gun.

The maneating continued, so the superintendent of police and two friends set up watch from a railway car on a siding near the station. Late in the night, when they believed no lions were around, all three men fell asleep. The maneater, which had been watching them, pawed open a sliding door and attacked the police superintendent. The other two men escaped, and a mob of frightened railway laborers held the door shut. The fearless lion escaped by jumping through a closed window, carrying the victim in his jaws. About a quarter mile away, the lion stopped to eat part of the body, and then moved on.

The maneaters were eventually shot by J. H. Patterson, and the years of terror finally came to an end. Some telegrams from station masters of that period have been preserved to remind us of their terror. One telegram reads: "Lion on platform. Please instruct guard and driver to enter carefully and without signal." Another reads: "Pointsman up telegraph pole by water tank. Train to halt, pick him up and proceed."

It has been estimated that a healthy maneating lion needs at least 50 people per year to survive, and about 150 people if he is to remain in prime condition. In Zambia, for example, one maneater ate 14 people within one month. And in northern Mozambique, Austin Roberts and Vaughan Kirby killed four maneaters which were killing 20 people per month. C. J. P. Ionides shed some light on the lion's appetite by reporting that one lion devoured a fat woman, and then had sufficient room to eat a wart hog—"for dessert" as he so wryly described it. Individual lions have been known to eat as many as three people in one night!

Few victims have been fortunate enough to survive the lion's powerful jaws. One was missionary explorer David Livingstone, who narrowly escaped death when attacked by a lion in Bechuanaland. Livingstone wrote that as the lion was "growling horribly close to my ear, he shook me as a terrier does a rat. The shock produced a stupor . . . a sort of dreaminess in which there was no sense of pain or feeling of terror." Most survivors report similar feelings, but there have been some cases where the victim felt "terrible pain."

Not all lions are maneaters, however. Most lions live their entire lives viewing man with complete indifference, and sometimes with fear.

Lion social life revolves around the pride. Prides generally consist of one to four males and 1 to 15 females, plus their assortment of cubs. In the Serengeti National Park (Tanzania), pride sizes vary anywhere from four to 22. The size of the pride is kept stable by death and by emigration of adolescent males, and less often, adolescent females.

It is the adult females who form the stable nucleus of the pride. Most lionesses remain together for their lifetime. Their knowledge of the pride's territory, the best hunting sites, and good cover is passed on from generation to generation of females. Dr. George B. Schaller, who studied lions in the Serengeti National Park for several years, has shown that the importance of the male's role is to help maintain the security of the pride's territory, thereby providing the females with a safe place in which to raise cubs.

We have often watched males patrol the pride territory, marking it with a mixture of scent and urine. The following is extracted from our field notes of early morning, March 11, 1973:

> "Black Mane and Rusty Mane are leaving the northeast shore of the soda lake, walking across the plain toward Ngoitokitok spring. Every 100 to 120 yards one male stops over a clump of grass, sometimes obviously sniffing it or rubbing his face in it. Then he takes two steps forward, scrapes each hind foot two or three times, raises his tail straight upright, and squirts urine. Sometimes both males mark the same spot, but more often they alternate."

Marking by males serves as a signpost to other lions that the territory is occupied. The strength of the odor is an indication of how recently it was marked.

The lion's roar is a second means of showing possession of a territory. The volume is sufficient to be heard several miles away. A roar begins first with one or two moans, followed by several full-throated roars, which then fade away into hoarse grunts. Roars also signal an individual's location to other pride members.

Lion life is a lazy life. Nearly 21 hours of every day are spent sleeping, dozing, or sitting. Because lions need to drink only once every few days, there is less need for travel. Usually they cover six to eight miles per day in hunting and patrolling their territory.

Both lion and lioness reach sexual maturity at about three-and-a-half to four years of age. When a female comes into heat, she and a male go off by themselves. Heat lasts three to five days, and mating occurs at the impressive rate of once every 10 to 20 minutes during this period. It is not without purpose though. Repeated mating is essential for female cats, since it stimulates the release of eggs from the ovary.

Lions normally prey on a large spectrum of animals. They kill whatever is easiest to catch, which usually means the sick or the careless. Sometimes they scavenge hyena kills. Some of the most important prey animals in East Africa are wildebeest, gazelle, hartebeest, zebra, wart hogs, impala, and Cape buffalo. Less frequent prey are porcupine, frogs, black rhino, hippo, giraffe, and ostrich eggs. Every adult lion in a pride probably kills 20 to 30 animals per year. Schaller estimated that every adult lion in the Serengeti National Park needs 6000 to 8400 pounds of prey per year.

In the realm of the lion, life is harsh. Most cubs are doomed to an early death. Three-quarters of adult male lions die in violence and agony: in poachers' snares, from hunters' bullets and in fights with other lions. Most of the remaining adult lions and lionesses eventually are expelled from the pride, and die in solitude, either from starvation or by being torn apart by hyenas or wild dogs.

They're out to Save the Black Rhino

by George W. Frame

A mother rhino and calf puffed and snorted furiously as they thundered across the plain within Ngorongoro Crater. Herds of wildebeest and gazelle hurried to get out of their way. Barely three yards behind the rhinos, a Land Rover bounced in hot pursuit. Game biologist John Goddard leaned out of the window and took careful aim at the adult with a Cap-chur gun. As the 10-cc syringe sped home, penetrating her rump, the 2,500-pound rhino reacted not at all.

"It's a hit," I shouted. "It injected!"

The rhinos continued to gallop at nearly 30 mph; it would be 10 minutes before the drug had a noticeable effect. John quickly loaded another syringe into his rifle and aimed carefully at the calf, which followed closely behind its mother.

"You got it, John! Driver, *endesha pole pole.*" These were the much awaited words, signifying that both mother and calf had received their carefully measured proportions of anesthetic. We now had a few minutes to catch our breath and hastily make notes while the Tanzanian driver, Stephen Ngereza, followed the two rhinos.

Darting and tagging are an important part of John Goddard's ecological and ethological studies of the black rhinoceros *(Diceros bicornis* L.) in East Africa. The metal ear-tags provide permanent identifying marks.

Black rhinos are reputed to be the most aggressive of the five living rhino species. The individuals in our study areas seldom let us forget this reputation, so we have to resort to immobilizing several of them to get data.

When a rhino is immobilized, we can take its body measurements, record the physiological response to the anesthetic, and collect parasitic ticks from the folds of its skin. We can pry open the drugged rhino's mouth to look at its teeth, to try to get an idea of its age. The information will help us develop a better understanding of whether rhino populations are stable, increasing or decreasing. Knowledge of the rhinos' well being under various conditions is essential in planning and managing reserves to ensure the survival of the species.

For centuries the three Asian species and two African species of rhinoceroses have been ruthlessly hunted, usually for their horns. Unlike horns of other animals, rhinoceros horns are entirely an outgrowth of the skin. They consist of a keratinous, or horny, fibrous substance, which is chemically identical to the fingernails and hair of man.

There is a widespread belief throughout Asia today that rhino horns have aphrodisiac value. This, coupled with the belief that various parts of the rhinoceros are useful for curing specific ills, has taken a heavy toll of the three Asian rhino species. Their future is very bleak.

The white rhinoceros and the black rhinoceros of Africa are endangered, too. As the rhinos of Asia become more scarce, the horn hunters attack the relatively more plentiful supply in Africa. Fortunately, the African rhinos are still numerous and if protective measures are taken now they may survive. Studies such as ours, hopefully, will lead to the development of sound management programs.

For months I had assisted John Goddard in his rhino study at Ngorongoro caldera and in the alternate study area at Oldupai (sometimes spelled "Olduvai") Gorge, 30 miles west of the caldera. Both are in Tanzania.

Comparison of the rhinoceros data from the forest and open-grassland/seasonal-marsh habitats of Ngorongoro with that of the dry-thornbush habitat of Oldupai would provide valuable insight into the needs and preferences of the black rhinoceros. But in order to accomplish a study such as this, it is imperative to be able to recognize every rhino as an individual.

Binoculars and a camera are essential tools. Normally, one approaches to within 20 to 40 yards of an individual rhino, either on foot or in a Land Rover. At this distance we clearly can see the minutest details, even the hair distribution on the animal's ears.

In thick bush the best method for identifying a rhino is to climb a tree and call with a mewing sound, like that made by rhino calves. The rhino will often walk right up to the tree.

Photographs and detailed written descriptions of every rhinoceros that we observed were used to compile an identification book for all the rhinos in both study areas. Key criteria in identifying an individual rhino are geographical location, sex, horn shape, torn ears, hair fringe of the ears, tail tassel and body scars. In several difficult cases we compared the facial wrinkles. Ear tags provide a long-term check on identification.

The adult black rhinoceros is essentially a solitary animal. Males are normally seen alone, and the female, too, prefers seclusion with her calf.

However, an immature rhino, which is driven away when its mother has a new calf, seeks the companionship of another rhino. Often it will attach itself to another immature rhino, or to an adult female.

"The largest group of rhinoceroses that I have seen together consisted of 13 animals," John Goddard reported, "but the group disbanded after two hours into solitary animals and groups of two and three. But I saw this large a group only once in my three years of daily observations."

The peak activity periods for rhinos are the hour beginning at dawn and the hour at dusk. At these times nearly all the rhinos we observed were actively feeding or walking.

Around midday most rhinos are asleep in the shade of a tree or else in a dust bowl in the hot sun. There daytime slumbers are sometimes interrupted for a snack or two on the nearest appetizing vegetation.

Our nighttime observations suggest that most of the population is active through the hours of darkness, though they do sleep part of the night. We always used a Land Rover at night; it would have been dangerous for a person to wander about rhino country in the darkness.

Different rhinos had different reactions to the vehicle's headlights. On the plains to the west of Oldupai Gorge I came across a mother and calf lying down. They were startled, and immediately arose. In confusion, both spun around several times with great agility. I kept the Land Rover in gear, ready to speed away if they decided to charge. The mother jabbed the air with her horn. Finally, after eight or 10 seconds of indecisive action, they ran off together away from the light.

Rhinos within Ngorongoro Crater seldom were frightened by headlights. Often they showed no reaction other than slight annoyance.

The black rhinoceros is mainly a browser, and has a strong upper lip to accommodate this feeding. I have watched a rhinoceros use its curved anterior horn to reach up and break off a four-inch thick branch of *Euphorbia* by pulling downward on it. The animal repeated this behavior several times. I suspect that rhinos with missing or broken anterior horns may have broken their horns on branches while feeding, rather than while fighting as is believed.

In Oldupai Gorge I frequently resorted to climbing high into the branches of these flat-topped trees. One reason was to get a better view of the rhinos feeding in the thick vegetation of the gorge.

The other reason was that I was chased there. Stalking a

2,500-pound rhino in thick bush is a challenging and nerve-wracking experience. One must be silent and stay downwind, listening for the chewing sound as the rhino masticates coarse acacia.

Sometimes my best efforts ended in failure. Then, I found myself running furiously for the nearest thorn tree, with a snorting "*faru mkali*" rhino in hot pursuit. I look back upon such experiences with fond memories, now that the thorn scratches have healed.

Rhinos travel well-worn paths. These trails and the regularity of daily activities are the main reasons that rhinos are so vulnerable to poachers. It matters not whether the weapon is a gun, spear, poison arrow or snare. The poacher, by knowing the hours of rhino activity and by using rhino trails, is assured success. He kills the animal, chops off the horns with his panga (heavy bush knife), and leaves the carcass for whatever may come along.

Predators sometimes attack rhinos, but these occurrences are rare.

In one instance, a male lion attacked an 11-month old calf. Felicia, the calf's mother, has a beautiful, straight horn that points forward like a saber. Felicia saw the approaching lion and prepared for the attack, while her calf snuggled closely against her. But then the calf panicked, and ran away with the lion in close pursuit.

Felicia trotted after the lion, who promptly diverted his attack from the calf to the mother. His jaws closed on Felicia's hind leg, and he clawed at her thigh. She whirled around with incredible agility, using her horn to stab the attacker in the ribs, neck and jaw, killing him instantly.

Lerai Forest covers one square mile of the floor within Ngorongoro Crater. Tall majestic acacia trees with dense bushy undergrowth ensure the availability of food throughout the year. The result is an extremely high density of 23 rhinos per square mile.

There are few water holes at Oldupai Gorge and palatable vegetation is less abundant. Consequently, the average home range of a single rhino covers about 12 square miles.

The grasslands within the caldera of Ngorongoro are intermediate in that they have less available food and water than the forest, but not nearly so little as exists at the gorge. The average home range in the grasslands is six square miles. This is only half as large as at Oldupai, but much greater than for the forest.

Individual home ranges are well defined, but overlap considerably—even among adult males. Rhinos sharing common parts of their range frequently come into contact with each other during their daily activities. Usually they are not aggressive, but confrontations do occur. One confrontation between two adult male rhinos remains particularly vivid to me. Horace is a docile old fellow whose home range contains grassland and swamp on the Crater floor. One day a strange male came down from the Crater wall and entered Horace's territory. Horace responded by attacking the intruder with hideous snorts and groans.

Both animals stood facing each other. Heads were lowered, ears flattened and tails raised. The intruder did not respond to Horace's noises, but silently stood his ground. Anterior horns were mere inches from each other; both jabbed and clubbed at the sides of each other's head. But actual physical contact seldom occurred. As usually happens, the intruder made a sudden retreat, closely pursued by Horace, the successful defender.

The black rhino's hearing ability is quite keen—so is its sense of smell. But rhino eyesight is notoriously poor. This became all too obvious to me one day on the dry, sandy windswept plains around Oldupai Gorge. After spotting two feeding rhinos with binoculars from two or three miles away, the driver and I approached very slowly in the Land Rover from downwind. As we moved closer, to within 150 yards, I left the vehicle and began to creep ahead on foot. Cautiously, I moved forward, a few steps at a time. When I found myself within 10 yards of the nearer rhino, I decided that that was close enough. Perhaps I was pushing my luck a bit too far. Despite my complete visibility both rhinos were totally unaware of my presence. The strong, gusty, noisy wind prevented them from hearing or smelling me. After one hour of observing their feeding behavior, I slowly backed away.

At Ngorongoro and Oldupai an average of four years is required for each adult female to recruit one calf into the population.

Sexual maturity is reached close to the fifth or sixth year, and a black rhino may live for 30 to 40 years. The population of black rhinos in Ngorongoro is at least 110, and we have identified 74 at Oldupai.

In the caldera, about seven calves are born per year, and at Oldupai Gorge only five calves appear yearly. Considering the occasional predation by hyenas or lions, and the usual population loss from old age, there is very little "margin of safety" remaining. Rhino populations can be exterminated by merely removing a relatively small percentage each year.

The white rhinoceros of Africa is much rarer. Poaching was once considered rampant in the Congo, where as many as 80 percent to 90 percent of the white rhinos were estimated to have been killed during the six long years of political instability. In spite of this their plight is not nearly so serious as that of the three Asian species: The Indian rhino is estimated to number about 700 individuals, and Sumatran rhinos may not number more than 180. The nearly extinct Javan rhino is known to exist only in Indonesia. Approximately 25 of this species are believed to remain.

In Vietnam the few remaining rhinoceroses are being exposed by the jungle defoliation program of the U.S. Army. Rhinos deprived of cover are very quickly eliminated, as are many other forms of wildlife. Saigon traders in recent years spoke of being able to sell a large horn for the equivalent of $2,000 (U.S.).

The future for the rhino is dim indeed, except in the few African countries such as the Republic of the Congo, Uganda,

Tanzania, Kenya, Zambia, and southern Africa, where the governments are taking steps to ensure their preservation. These governments, and those of us who are studying the rhinoceros, agree that only in national parks like Tsavo and conservation areas like Ngorongoro Crater does the rhino have a chance for survival.

The Man Who Makes Pets of Gorillas

by Dick Bothwell and David Coleman

This giant weighs as much as three good-sized football players, puts dents in cars with irreverent nonchalance and scares the daylights out of practically everybody. Yet, looking like an enormous, hairy Buddha, this gorilla, Tommy, has a side to him that many people would not suspect—he's a gentle, even loveable pet.

Robert E. Noell, who runs Noell's Ark Chimpanzee Farm near Tarpon Springs, Fla., has been raising gorillas for almost 20 years. He knows from experience that these anthropoids (meaning man-like) are neither as ferocious nor as hardy as they look.

"We got our first one in 1950," says Noell. "She only lived a year. Then we bought a male; he lived a year. The third one lived four or five years. . . ."

Noell's Ark lost two more after only a few years (each pet costs about $5,000). But they have managed to keep Tommy for the past 10 years.

Because of their high price and tendency to catch colds and more serious ailments easily, there aren't many "pet" gorillas in the U.S. Tommy is one of the few privately-owned ones, and it's been estimated that there are only about 85 in captivity in the whole country.

People just aren't used to seeing 600-pound King Kongs walking around, and the frenzy of activity that follows one of

Tommy's unannounced appearances rivals the slapstick comedy of the Keystone Cops.

The day he waddled calmly into the midst of a group of railroad workers near Tarpon Springs is still cause for a string of frenetic adjectives from stammering workers who couldn't believe their eyes and who involuntarily joined in the fad of seeing how many people could simultaneously cram into the nearest parked car.

Noell's Ark often tours the country, and one of the attractions of his troupe is his wrestling matches with the giant ape. The anima.'s huge, black hands palm the man's head as a pitcher would palm a softball. Part of the routine includes Tommy putting Noell's arm between his jaws in mock bites.

"When we're on the road," he says, "I go into the cage with Tommy each night. And I always tell the audience, 'This may be the last time.' But he's gentle; we brought him up like a child."

But this "child" can be extremely dangerous all the same. The largest of the anthropoid apes, gorillas weigh around 450 pounds at maturity and reach a standing height of about 5½ feet. Tommy weighs considerably more than a wild gorilla due to lack of exercise and a diet of two bushels of fruit and vegetables every day—and he won't be fully grown for two more years.

Gorillas have moods of irritation and anger just as humans do. They are the most dangerous during these periods because they lack the restraint and discipline of humans—behaving much the same as a criminally insane person.

However, Noell has raised Tommy from an infant. The animal has been conditioned over the years and cringes when his owner speaks sharply and raises a hand in threat.

The ape is so well-trained that Noell can even share a banana with him mouth-to-mouth—it's all part of the act.

Theatrical as it seems, Noell has accomplished a remarkable job of training the beast. Gorillas—unlike chimpanzees and orangutans—do not learn easily. Although gorillas have shown an ability to solve problems involving the use of tools such as sticks and stacked boxes to reach suspended food, they are ". . .

slow in adaptation and limited in initiative, originality and insight" according to studies conducted by the American psychologist R. M. Yerkes.

The owner's method of reward, praise and punishment has resulted in a *relatively* well-behaved pet. The reward part of the training is a little rough on Noell's wardrobe, however. He keeps candy in his shirt pocket, and the ape knows it. The removal of candy along with the whole pocket is a daily occurrence. Noell's shirts come from Goodwill Stores by the dozen at 15 cents each.

Known technically as *Gorilla beringei*, gorillas such as Tommy come from the high mountains of the eastern Belgian Congo. A similar species *(Gorilla gorilla)* thrives in the forest regions of West Africa. Its main difference is a much thinner coat of fur than the eastern form.

These anthropoids are docile vegetarians. They won't attack unless severely provoked or threatened; but when their anger *is* aroused, they become extremely dangerous aggressors. Tommy, if angry or afraid, has been known to dash about attacking anything in sight. One of his casual swats has been known to put a $50 dent in Noell's parked car.

But Noell is not discouraged by Tommy's behavior. In fact, he has another "pet" like Tommy growing up right now—Tarpie, a three-year-old, 60-pound female.

Marmosets: The World's Smallest Monkeys

by Barbara Ford

Mrs. Germaine Miller, a tall, fashionably-dressed blonde, boarded the boat train at Paris a few years ago for the trip to the English Channel. She carried two purses, one of expensive leather and the other a small wicker affair. As the train was rolling through the French countryside, the passengers heard a shrill chirping noise. "There must be a cricket in here," exclaimed someone. Everyone looked around, but no insect was seen. The chirping noise continued.

Later, after she and her fellow passengers had passed through customs, Mrs. Miller offered to exhibit the "cricket." She opened the wicker basket. Inside was a tiny, brown, furry creature with a long tail and an almost human face.

It was a monkey, but a special kind of monkey. Mrs. Miller, who lives in Westchester County, N.Y., raises pygmy marmosets, the smallest monkey in the world. The animals are, in fact, the smallest living primate. Excluding their tails, they average four inches in length in the wild and an inch or so longer in captivity. They weigh only a few ounces.

An animal this size is easy to carry around, and Mrs. Miller and her husband, Roswell, tote one or more pygmies (they have three at present) with them almost everywhere they go. The animals have been on airplanes, trains and ships, and in cars, beauty shops, restaurants and stores. Chou-Chou, a five-year-old female, has visited Europe, Canada and Puerto Rico, as well as Florida and the western states.

"People don't know we have a monkey with us—they just think I'm crazy for carrying two purses," says Mrs. Miller.

So far as she knows, Mrs. Miller is the only person in the country raising pygmy marmosets. Delicate and high strung, the little creatures are hard to keep alive. Even larger marmosets are difficult to raise. Irving Levane of the New York Simian Society, an organization of monkey owners, knows of no marmoset owners in the area, although he estimates there are 10,000 to 20,000 owners of other kinds of monkeys in Greater New York. Most pet store owners have stopped stocking marmosets. "We come in in the morning and find them dead in their cages," says one animal dealer.

What makes marmosets so hard to raise?

According to Dick Bergman, who's in charge of the monkey house at the Bronx Zoo, "South American monkeys like the marmosets are always harder to keep healthy. If they don't get extra calcium, and Vitamin D, they go down with rickets, TB or colds." With the right diet, however, Bergman thinks the marmoset has a good chance of survival in captivity. The "right diet" would make most people pause. Besides fruit, insects and special formulas brewed by Bergman in the zoo kitchen, it includes mealworms flown in from California—an absolute must in the marmoset diet.

The diet works, though. The zoo's common marmoset died recently at the age of 14, which is about as old as a marmoset has ever lived to be in captivity. At present, there are 10 marmosets in Bergman's charge—seven white-faced, one cotton-topped, and two pygmies—all of them in good health.

Not everyone would like the marmoset as a pet even if it could be kept alive easily, Bergman cautions. Marmosets are shy, nervous monkeys, likely to bite when excited (Mrs. Miller has been bitten so many times she has lost count). Owners may find their pet less than enchanting when it sinks all 36 of its needle-sharp little teeth into their hand. "That's when we get all our 'donations'," smiles Bergman.

Mrs. Miller expresses some bewilderment about her unique success. "People are always asking me what I do; my marmosets are so healthy. I don't know." But a description of her daily routine is revealing. Before the pygmies go outside in cold weather, she dresses them in minute sweaters, which she knitted herself, to protect their lungs. In the home, the animals never run around on the floor. "Never," she cries, exhibiting tiny chains made out of a number of inexpensive children's bracelets by which the monkeys are tethered to rings in the windowsill.

At dinnertime, the creatures perch on Mrs. Miller's dining room chair and eat bits of food she presents to them. "I give them tidbits—they eat their mealworms themselves." The "tidbits" vary with the meal. On the French Line, the well-traveled Chou-Chou developed a fondness for lobster and ice cream. Like Dick Bergman, Mrs. Miller improves her pets' diet with food supplements combined in a formula she concocts herself.

Sick marmosets visit Mrs. Miller's personal physician, not a veterinarian. Mrs. Miller doesn't have a high opinion of veterinarians. "They don't know anything about monkeys," she says. "Dogs, cats, perhaps, but not monkeys." She travels with a medicine chest, not for herself, but for the monkeys.

Her pygmies are exceedingly well-trained. "Do pee-pee, do pee-pee," she tells Chou-Chou, presenting her with a paper napkin, and Chou-Chou obligingly does so. "One of them sleeps with us, but it goes to the toilet every night before it goes to bed, so there's no problem," she remarks. The marmosets also have a "trick." "Give me a kiss, Coco," she commands, holding him in her hand. The animal bends forward and pecks her nose with its lips.

Mrs. Miller estimates that she spends "all her time for three months" with each animal after it's born. The procedure obviously pays off. In 10 years, she's had some 15 marmosets, most of which have lived to respectable age with her. One marmoset acquired from a pet shop as an adult lived with Mrs. Miller for over seven years before it died. One of the current pygmies is five. The others—twins like most marmosets—are three. Even more impressive, perhaps, is the fact that her animals have produced 12 young, most of which have survived.

Does Mrs. Miller recommend marmosets for the average pet lover? "No!" she says. "People don't know how to take care of monkeys. Every marmoset but one that I've given away has died."

In the past, however, marmosets were very popular as pets. Sixteenth century voyagers from Spain and Portugal saw the tiny animals being kept as pets in the Amazon region, and brought them back to Europe. They soon became a fad among the aristocracy. Women were fond of them, carrying the furry mites around in their capacious sleeves. The name "marmoset" is believed to come from the Old French word "marmouset," which means a grotesque figure or manikin.

The marmosets taken back to Europe by these early explorers probably included a variety of the 30 or more species of marmosets. The pygmy is the smallest, but all of the marmosets are small—collectively, they're the smallest monkeys in the world. Because of their size and dense fur, observers who spot them from a distance often mistake them for a kitten or a squirrel. Up close, however, a marmoset's expressive simian features mark it as a monkey.

If you approach the white-faced marmosets at the Bronx Zoo, they'll dart as far away from you as possible, raising their brows and contracting their lips in an expression of alarm. When Mrs. Miller pulls a sweater over the head of one of her pygmies, it contorts its face with rage while it chatters angrily.

The marmoset's hands also mark it as a monkey, but a primitive one. Fingers move in the same plane, so that it must press an object against the heel of its palm to hold it. Each of the long fingers is provided with claws rather than nails, a peculiarity that distinguishes it from all other American monkeys.

All marmosets share these traits, but the colors of their coats and other superficial features divide them into a number of species. Exactly how many is unknown. A few may simply be local varieties. Marmosets exhibit an astonishing range of colors and patterns in their coats, as well as a spectacular assortment of tufts, plumes, manes and even mustaches. The grandest-looking is probably the lion-headed marmoset, one species of which has bright golden fur over its entire body, including the lion-like mane. Other species display patches of golden fur. A second candidate for the most outstanding marmoset honors is the aptly-named white shouldered marmoset, which has a white head, chest and face, as well as white ear tufts. The silver marmoset is all white, a pure, silvery color that seems more appropriate for the Arctic than the jungle.

One imperious-looking creature, the emperor marmoset, has, around its mouth, long white hairs which fall to its shoulders like an overgrown mustache. The plumes of white hair on top of the head of the cotton-topped marmoset make it resemble a Masai warrior. A number of marmosets boast eartufts either in the form of long, drooping hair that arises in front of the ears or corollas of stiff hairs that grow all around the ears.

When a marmoset is angry, this abundant fur stands out—an impressive sight even in so tiny a creature. If you approach Mrs. Miller's Chou-Chou when it's eating one of its favorite treats, a miniature marshmallow, the little animal will cram the marshmallow in its mouth, lower its brow and literally swell with rage.

Later, if Chou-Chou lets you touch its fur (wild marmosets dislike being handled, but the Miller marmosets have been trained to submit to it) you'll find it's almost as soft as down. The fur of even the larger marmosets is soft and silky.

With coats like this, it's not surprising that some authorities consider marmosets the most beautiful of the monkeys. But not all marmosets are attractive. The silver marmoset has a gorgeous white coat, but its ears and face are naked pink skin with red blotches. The top of its head is bald, too. Quite a few marmosets are bald. Martin's Marmoset has a bald head described by one naturalist as being "a sickly puce." The baldness is not the result of falling hair; bald marmosets are quite simply born that way.

For a small creature, the marmoset makes an enormous amount of noise. To the uninitiated, these cries often sound like those of insects or birds rather than monkeys. But when it wants to, the marmoset can make these shrill sounds carry for some distance. When one of the Miller marmosets is in the first floor living room, it can easily make itself heard in a second floor bedroom of the large (30 rooms) house.

Marmosets have a whole repertoire of noises. According to one authority, the black pencilled marmoset has different cries for contempt, anger, hunger, fear and even boredom (pseeeh, pseeeh). Another naturalist claims that marmosets have a "limitless number" of calls, many produced only in response to certain objects, such as a black dog or a cake of soap. Mrs. Miller notes that her pygmy marmosets greet a black poodle belonging to her daughter with a cry that is reserved especially for that particular dog.

Marmosets not only utter cries in isolation; they talk to each other. "My Mimi used to call the roll of her children when we went on vacation," says Mrs. Miller. "And each one would answer." When Mrs. Miller and her husband drive in their car, one marmoset is usually perched on Mr. Miller's shoulder while the others occupy a closed basket. As the car nears the Miller home, the "outsider" calls to the monkeys in the basket, seemingly telling them they are near home, and the others answer.

For one reason or another, then, marmosets seem to be talking most of the time. "They're always chattering," says Dick Bergman. If you sit in the living room of the Millers' apartment in New York City, an almost constant chirping and trilling is heard from the kitchen, where two marmosets occupy separate baskets on the sink. We may not even hear all the sounds marmosets make. Sensitive equipment has recorded marmoset sounds which are too high for humans to hear.

Talkative, pretty and tiny, marmosets have an almost immediate appeal for anyone who comes across them. According to the experts, though, a marmoset is better off in the jungle or a zoo, unless you're fantastically patient.

The Horse of the Caveman

by Bruce Frisch

Hoots and cries and the thump of drums floated from the forest at the top of the cliff. A few stubby yellow horses trotted out of the trees, stamping nervously at the cliff's edge. As the din grew louder, more horses dashed out. Some bristled with feathered darts. When a circle of shouting, bearded cavemen burst out of the woods waving their arms, the milling herd surged over the cliff and crashed to the ground below.

That evening, 25,000 years ago, the darkness glowed with a score of fires over which the horses roasted. The feasting Stone Age men ate the carcasses right down to the bones. Then they cracked the bones for the marrow and tossed the pieces on enormous piles. Eventually, the bones of 100,000 horses and other animals formed a rampart around their summer encampment at what is now Solutré in southern France.

The wild horses of the cavemen probably were the single most important ancestor of the modern horse, *Equus caballus.* Unknown to Europeans, they also survived as a separate species in the deserts of central Asia until discovered at the end of the last century and named Przewalskii's horses. They were the last truly wild horses. All other supposed wild horses, including our mustangs, are feral—domestic breeds gone astray.

Przewalskii's horses were immediately recognized as the spitting images of the horses drawn by Neolithic man on the walls of his caves. Running, falling through the air, streaming feathers like a stuck bull, the Roman-nosed little horse with no forelock and an upright mane was unmistakable.

Today, all that are left are about 152 animals in zoos, and desperate measures must be taken to prevent their extinction.

In Stone Age times, herds of many local varieties of wild horses were spread across Europe and northern Asia. They remained wholly wild until 2500 B.C., when nomads of the Asian plains tamed some for transport. In 2000 B.C. they reached civilization in China and Mesopotamia. The first arrivals were hitched to chariots, because they were too short and stocky for riding. Later the Arab bred the horse for size and speed and rode him on a wave of conquest. Under North African Berber horsemen, he galloped over Spain.

Meanwhile, in northern Europe, the horse had been bred for enormous size and brute strength to carry armored knights. A line of these tank-like beasts met and repelled the Moslem invasion in southern France.

While in Spain, the Barb horses of Africa mixed with local wild types, and it was this cross the Spaniards carried over the sea to conquer the New World. In February 1519, Hernan Cortes sailed from Havana and landed in Mexico with 17 horses. These were then the only horses in North America.

From cattle ranches of Spanish missions that later were established in the southwest United States, the Plains Indians got the horse and developed a new culture around it. The horse was perfectly suited to the area, because these plains of North America were his birthplace.

Sixty million years earlier, before the plains had turned to grass, Eohippus, "dawn horse," browsed the wooded river banks. Stupid, with toes and padded feet, the dawn horse was the size of a large dog. But even then, it was as fast as the modern horse. Over the next 35 million years its ancestors grew bigger and smarter.

When the climate changed, and the plains grew a blanket of

grass, the primitive horse eagerly exploited it. He became a grazer, with teeth designed for grinding and long wear. He grew hooves on the ends of his toes and sprouted to a height of 40 inches at the shoulder. This model of the horse was a "sensational success," says Dr. George Simpson, a paleontologist at the American Museum of Natural History, who has traced the history of the horse's evolution.

Whenever the land bridge across the Bering Strait was high and dry, the current version of the horse would cross over to Asia. The line always died out in Eurasia, but continued developing in North America. About one million years ago the modern genus, *Equus,* finally emerged and spread to the Old World. Whether *Equus* had also developed independently in Eurasia from a slightly earlier migrant is unknown.

The final development of the modern *species, Equus caballus,* took place in Eurasia. Most authorities believe *Equus caballus* is made up of a large measure of *Equus przewalskii* and smaller dashes of other stocks. A few scientists consider the Przewalskii just a local variety of *Equus caballus.*

At the time the Indians arrived in North America, wild horses were still here, and survived past the Ice Age. Then they suddenly and mysteriously disappeared, along with the mastodon, ground-sloth, direwolf, and various others.

In historic times, Western Europeans knew nothing of the wild horse or the cave drawings. Then, in 1869, a man hunting foxes stumbled across the entrance to a cave at Altamira in northern Spain. Ten years later, a local nobleman began digging it open. One day he took his five-year-old daughter in with him. While playing 30 yards inside the entrance, she looked up at the ceiling and cried, "Bulls!"

For two decades no learned authority would believe the drawings were by cavemen; they were dismissed as forgeries, perhaps a practical joke. But when another cave was found in southern France in 1895, and two more in one week in 1901, skeptics were won over. Many more caves have been found since, and there may still be some undiscovered. Perhaps the most famous, Lascaux, was found by four boys in 1940, and another turned up in 1956.

One of the two caves found in 1901, Les Combarelles, was painted with 116 horses. Naturalists were quick to notice the striking similarity between the big-bellied horse in the drawings and a new find that had just arrived in European zoos. These were called Przewalskii's horses after Colonel N. Przewalskii of the Russian Imperial Army. While traveling in Asia in 1881, the colonel had been presented a skin by a local official who had got it from men who hunted camels in the central Asian deserts.

In 1899–1902, 54 of the horses were captured and sent to Siberia, Poland and Russia. In 1901, Carl Hagenbeck, the famous animal supplier of Hamburg, sent an expedition to the eastern end of the Gobi desert in Mongolia, near where China, Russia and Mongolia meet. Two hundred Kirghiz tribesmen were recruited to round up the little horses. But the horses were so vicious and unyielding the collectors had to kill mares, take their foals and give them to Mongolian pony foster mothers.

"They're wild as hell," and like the zebra, would be harder to break than a mustang, says W. B. House, curator of mammals at New York's Bronx Zoo.

Of the 40 animals Hagenbeck captured, 32 went to the Duke of Bedford for his private zoo at Woburn Abbey.

The large herds of those times are gone. The last animal to be captured was taken in 1947. Only a few bands of five or six were seen by hunters in the early 1960s. Since then, it is believed, a succession of extremely severe winters may have wiped them out completely.

Only two places are left in the world with herds large enough to breed for survival, says House. At least 20 horses are needed. The Prague, Czechoslovakia, zoo has a herd numbering in the 30s; the Catskill Game Farm, south of Albany, New York, has 20-odd.

Formosan deer, Pere David deer and wisants, or European

bison, also exist only in zoos. "But God help us if we have to depend on zoos to perpetuate animals," fumes House. Among zoomen there has been, and pretty much still is, prestige in having a large number of species. As a result, a zoo gets one or two specimens of an animal and sells any offspring.

And "we mismanage our preserves," such as Yellowstone Park, continues House. Deer, for example, are fed hay to carry them over hard winters and are not allowed to be shot. They then become so numerous they strip their range, damaging it permanently. "I'm not optimistic about the future of wild animals," House sums up.

"We are reducing the number of species at this zoo and doing it right with the ones we have," says House. He has just bought three yearling Przewalskiis with which he hopes to build a herd of 15 or 20, then start another.

If he is successful, America may be the last, as well as the first, home of the wild horse.

The African Ostrich: Bird of a Different Feather

by George W. Frame

Diamonds were the cause of South West Africa's mammoth ostrich slaughter more than a century ago. It began when a hunter shot a wild ostrich and found several shiny gems in its gizzard.

Permits soon were issued and ostrich hunting became the popular pursuit of those seeking quick riches. The hunter-prospectors found many diamonds in the slaughtered birds—as many as 53 in just one. They also learned there was a big market for ostrich skins, and at the height of the "Great Hunt" 12,000 were exported during a five-month period. The killing eventually dwindled because too few birds were left to justify hunting efforts.

Man's use and abuse of the ostrich date back more than a century, however. Their eggshells were used as drinking cups almost 5,000 years ago in the Middle East. Roman and Egyptian noblewomen rode tame saddled ostriches on special ceremonial occasions, and Egyptians regarded the bird's feather as a symbol of justice because its vanes are of equal length on either side of the central quill.

The ostrich has been domesticated to do many things, but no story compares with one told by two South African farmers who trained a pair of ostriches to herd sheep. They had raised the birds from fledglings, which followed the herdboys as chicks. As they grew older they began taking an active part in herding and seemed to enjoy it. Allowed to exercise some authority, their self-importance grew to the point where they actually resented the herdboys' presence.

Eventually the farmers sent the sheep to graze under the sole direction of the ostriches. All went well. Promptly at dusk the herd returned with the birds fussing about them, pecking at the tails of stragglers. But one day the sheep were late, and a farmer rode out to learn why. He discovered one sheep had died and the ostriches were trying frantically to urge the motionless carcass to follow the other sheep.

The real incentive for domesticating ostriches came in 1838 when their feathers first were exported commercially from South Africa. During the following decade the use of ostrich products became widespread. Eggs were sold to housewives and bakers, and the fat was used in cooking. Some meat was eaten fresh but most, cut from the legs and back, was dried to make biltong. Even now a slaughterhouse operates in Oudtshoorn, processing about 480 birds each week.

Feathers again became fashionable in Europe and America during the 1880's and the ostrich industry once more boomed. When it collapsed in 1914, South Africa was exporting more than 3,600 tons of feathers annually. In the past 25 years the demand has resumed and 200 farmers now are raising more than 25,000 ostriches in the Oudtshoorn district.

At one time the birds roamed over almost the entire African continent, from the Atlas Mountains in the north to the Cape of Good Hope, skirting only the Congo forest, Zambia and Malawi. Fossils of bones and egg shells have been found across southern Eurasia to China. In fact, the birds remained in Syria and Arabia until at least the 1940s.

Ostriches are the largest living birds. Fully grown, they may stand eight feet tall and weight 300 pounds. Lacking ability to fly, they depend on their feet and long legs for locomotion and self defense. Their kick is a powerful weapon and they can direct it high enough to knock a man off a horse. It is doubly lethal because of a long nail on each toe. S. C. Cronwright-Schreiner, a farmer, wrote in 1897 that an enraged male ostrich "kicked a hole through a sheet of corrugated iron behind which a man had taken refuge."

Their feet also are specially adapted to running. The two toes on each foot and the padded soles work efficiently in crossing long tracts of yielding sand. One toe is much smaller in size, suggesting the ostrich may be evolving toward a one-toed foot like zebras and horses. When it runs the ostrich is a picture of both comedy and grace. The head and neck are held back and the wings are kept clear of the thighs like graceful fans. Only their legs seem to be moving in a rhythmic, smooth stride.

They can maintain a speed of 30 to 35 miles an hour for as long as 30 minutes without fatigue, and at full speed they have been known to jump a five-foot fence. Even a horse is no challenge to them. On horseback at full gallop, the African explorer Frederick Courteney Selous chased an ostrich over a mile but could not overtake it. The bird's long stride averages about 12 feet but can double that at top speed.

Although distinctly unbirdlike in their running ability, they do act like all birds in eating gravel to grind the food in their gizzards. Ostriches consume sand, stones, grit and bones for this purpose, and are particularly attracted to shiny objects, particularly glass, shiny pebbles, and those diamonds that had such fatal consequences. Zoos have recorded their appetite to include coins, jewelry, nails and penknives.

In studying the social behavior of the ostrich, Franz and Eleonore Sauer of the University of Florida found that troops and families of the birds often join together in peaceful communities of up to 600 members with individual groups remaining distinct. Outsiders can only make social contact by approaching the community in an appeasement posture, which consists of having the tail pointed vertically downward and the head inclined.

During the breeding season the male ostrich becomes vicious and unpredictable in temperament. A roar, or "brom," is uttered at this time. It consists of three booms—two shorts followed by one long—which reverberate across the plains. Roaring appears to be both a courting note to females and a challenge to other males. The males fight often, but seldom with fatal results. The large callous breast pad, which normally supports the body while resting, protects them from powerful kicks.

The Sauers learned that ostriches are usually polygamous if enough females are available. Normally every male has a principal female, and two subordinate females which he drives or lures away from the troop. These secondary females are tolerated by the primary one and all lay their eggs in the same nest. The principal female has prior hatching experience and often chases the secondary ones away from the nest when the eggs are laid.

I once saw their courtship ritual, which in many ways is as elegant as a peacock's. The male had wandered away to a secluded spot with one of the females and grazed with her for a while, a ceremony which precedes mating. Gradually the two birds synchronized their movements with increasing precision.

If perfect synchronization is not achieved the courtship proceeds no farther. In this case it was and I saw the male become increasingly excited. Kneeling on the ground and alternately flapping each wing, he rocked from side to side, twisting and turning his neck in a series of quick spiral movements. His head and neck were lowered until on a level with his back, and as each of the wings flapped he moved neck and head in that direction. The back of his head struck with a loud click against his ribs, first on one side and then the other.

The female's response was to lower her head, open and close her beak, and flap her wings. As the male called his dull roar she circled around him. When he jumped up she dropped to the ground and mating finally occurred with a tremendous flapping of wings.

After mating the male and female dig a shallow depression in the ground about three yards in diameter far from trees or shade. As the female lays each egg in front of her mate he tucks it under his body. Each female may lay anywhere from a half dozen to dozens of eggs. Their incubation is a task shared by the male and female although the female begins, remaining up to two days and nights before the male takes his first turn.

Besides man there are other animals interested in the eggs as food. Hyenas, bush pigs and jackals feed on them. Egyptian vultures are especially adept at cracking the tough shells by taking stones in their beaks and dropping them on the eggs.

Ostriches reach maturity at 18 months, when their color changes from a mottled brown to the black and white plumage of the male or the gray-brown of the adult female. The main enemies they must face are lions, leopards, cheetahs and hunting dogs.

Besides running and kicking, ostriches also resort to an open mouth threat gesture when confronted by a creature large enough to be dangerous. This is accompanied by a hissing note for a mild threat and progresses to loud snorting and finally loud tonal calls as the bird becomes more excited.

Bernard Grzimek, the German zoologist, also noted that fleeing ostriches sometimes disappear abruptly before reaching the horizon. They do so by sitting with their long necks flat on the ground. This most likely is the origin of the legend that the ostrich hides its head in the sand when frightened.

Even with its evasive techniques the ostrich was almost unable to escape the threat of extinction. Realizing this, the Cape Colony in May 1870 passed a law to conserve wild ostriches by requiring a license to kill, capture, hunt or wound any wild specimen. Prior to this, wild ostriches were classified as vermin, along with lions and leopards, because they damaged crops.

When their feathers went out of fashion and became quite cheap during World War I, public sentiment turned against the birds and hunting restrictions were eased. Ostriches were chased in cars and slaughtered wholesale. Hunters often returned from a single trip with 400 to 500 skins. The meat and feathers were left lying on the ground.

Ostrich farms helped curb some of this killing by reducing the need to hunt the birds, and today African national parks and game reserves are effectively safeguarding the species.

Larger birds than the ostrich—the giant moas of New Zealand and the elephant birds of Madagascar—were hunted by man as recently as 700 years ago. Now they are extinct, probably because too many were killed and too many of their eggs stolen. The ostrich came close to the same fate, but protection afforded by domestication, national parks and game reserves has now given it a new future in eastern and southern Africa.

How We Found Mexico's 'Extinct' Tortoises

by Ray Pawley

People generally think that discoveries of new animals are a thing of the past. Today, it seems hardly possible that any sizable new creature could have escaped notice. Well, it can still happen. I was fortunate enough, two years ago, to be a member of a field group that did stumble on living specimens of a creature that, until a few years ago, had been considered as extinct as *Tyrannosaurus rex*—a creature whose existence, oddly, may be partially due to mankind's benevolent attitude toward it!

In late 1965, the Chicago Zoological Park, with assistance from Ted Borak, an enterprising automobile businessman, sent an expedition to Mexico consisting of myself and photographer Alan Levine. Our purpose was to gather a series of vampire bats for a new zoo exhibit. We were one of the fortunate few who were granted limited permission by—and help from—Dr. Hernandez Corzo, director-general of the Department of Conservation of Mexico, to investigate various portions of Mexico with respect to collecting some zoological specimens for exhibit and educational purposes.

A few miles from the nearest highway, on top of a small mountain, we looked across an extinct lake bed in the northern Mexico desert country. Although it was difficult to envision a vast lake at this site, we could see some evidence that perhaps our desert was "younger" than appeared. For one thing, a vanishing portion of a lake was still in view, miles away, and our vantage point may have been a shore-line a few eons ago. The characteristic giant cactus or agave of desert regions was conspicuously absent in the dry lake floor. A part-time archaeologist in Torreon had told us of Indian caves in these areas that contained quantities of discarded shellfish remains.

Later that afternoon, we discovered a striking number of large holes, or burrows, dug into the ground, all proceeding back into the hillside at a gentle slope. The amazing thing was that the entrances to these burrows were clustered in an area of half an acre or less. Though we saw no signs of the animals who made them, natives told us that they were the houses of big turtles which they called "tortugas."

The concentration of lairs seemed peculiar. North American tortoise species are independent and scatter with their burrow sites. What kind of big tortoise would have this strong colonial tendency? Oddly, too, the burrows were not dug on the dry lake floor, nor up on the side of the hill or mountain. They were dug just above a point where the hillside *meets* the lake floor. From this location the tortoises would have to travel many hundreds of yards to the nearest food.

Did they avoid building their town on the desert floor because some ancient instinct taught them to avoid areas of potential flooding? Was their "town" so old that through the many years their natural forage had changed so that they would have to travel further from home for a meal? Why weren't there any smaller holes or burrows such as small tortoises would use? Did the young live in the adult tortoise burrows (very doubtful), or was this colony dying out? If so, was it a unique kind of tortoise not described in the books we'd read?

Since we saw no tortoises venturing about in the midday heat, we decided we would have to *dig* after one! As it turned out, this was no easy matter. The burrow was not simple and straight like that of the gopher tortoise of the eastern U.S.A., but it was branched. Every four feet or so, the tunnel would fork and re-fork, with a total of about 22 feet of tunnel work, up to six feet or more underground. It was at the terminal end of the largest fork that we unearthed our prize . . . a huge fellow with predominantly yellow color (including the iris of the eye). Its huge size (like an "overnight bag" in volume) was unmistakable. It had to be the "Tortuga Grande" or Torreon tortoise—long "extinct" according to old literature.

As "new" animals of the Americas go, invertebrates are still being discovered and catalogued at a somewhat regular rate. Vertebrates, however, are fewer and relatively more conspicuous. In the more impenetrable tropics, new frogs and lizards as well as new birds are infrequently found, but few new mammals are being discovered. The slow-moving, highly obvious tortoise is the most unlikely candidate of all.

The New World is surprisingly short of tortoises, particularly north of the Panama Canal. The original three species of tortoise from North America (members of the genus *Gopherus*) have been known to science for more than 100 years. In spite of all this, the big Torreon tortoise (up to 10 pounds or more, and well over 16 inches long) did elude the gaze of science until only a few years ago.

Suspicion of this tortoise's existence had become known, we discovered, when two men from the University of Illinois discovered two surprisingly large, empty tortoise shells (carapaces) in a trash dump of a small town, Carrillo, Mexico, in August of 1958. Herpetologist John Legler visited the same desert area in September 1958, and resolved any remaining doubts when he collected a number of live specimens for study. His successful efforts led to his publication of a description of the new species in April 1959, bringing this fourth and largest species of North American tortoise to the attention of the scientific world. So—unfortunately—we weren't the first discoverers, but we were close on his heels.

How could something so conspicuous, slow-moving and widespread escape the detection of science so succsssfully? They are found not more than 100 miles south of our U.S.-Mexico border, and at least one heavily traveled highway cuts through a large portion of their natural range.

If the tortoise were secretive, nocturnal or hiding in a jungle, it wouldn't be so embarrassing. But, no, this ponderous animal wanders about in morning or evening in full view of highway traffic. How many more undescribed species of animals remain to be seen . . . maybe in the trash heaps of our *own* back yards?

A partial explanation may be that this tortoise is exceedingly well-known within its own natural range, but the local people never thought about it being "unknown" since it was so common! And, scientists don't always go poking around deserts asking for tortoises! Thus, due to an amusing divergence in circumstances, the two camps never made contact on the matter.

If these tortugas lived in the region at the time the lakes were at full level and the area was green, it is easy to see why this would be a choice tortoise site. However, if they successfully underwent the necessary behavior changes as their environment might have changed, the adaptability of this species is astounding. Most animals migrate to new areas the moment their environment changes even the slightest bit. But to adapt from a lake area to a desert area is nearly impossible. If this is so (we still don't know), their changes have been most ingenious.

Instead of staying "topside" during the day, or most of it, these tortoises now limit their excursions to a few hours in the morning or evening to escape the scorching daytime heat or freezing night cold. Since they are remarkably active as tortoises go, they may make better use of the short available time in feeding. In fact, because tortoises can go a long time without food, they remain dormant for months at a time in the summer and winter, emerging only during a few short rainy evenings in a year!

Their ability to retain body moisture while living in their cool, dark burrows is also understandable, particularly as their skin and shell are almost impervious to moisture seepage. In fact, as do most other species of tortoises, they apparently derive their necessary moisture from the green plants they eat, storing water in a sizable reservoir in their own body for use in dry spells.

One gets the impression he is looking at perhaps the last remnants of an already relict species that should have been extinct before now. But how have they managed to "hang on" so long?

The answer may be man's efforts, and largely on a localized, uneducated level to boot!

From evidence exhibited in trash heaps dating back beyond the present Tarahumara Indians, these reptiles have long been relished as an article of diet by man. Yet, further back than anyone can remember, men would take loads of greens to the tortoise colonies, and feed the animals at periodic intervals—a fundamental practice of semi-domestication.

Nowadays, every week or so, someone will travel around with a pickup truck-load of alfalfa and drop it off at the tortuga colonies. Reportedly, the tortoises come immediately to the surface and begin grazing, even before the truck leaves the area. Here, then, is a working system of careful cultivation and cropping of an animal species that has been going on for no one knows how long. It may even be conceivable that some of the colonies would have died out except for this haphazard but helpful practice.

Much more work remains to be done with this intriguing animal. Despite the "welfare program," there is every evidence that the tortuga is on the way out. The game laws of Mexico protect the species, but it remains for research to take over, and fill in the broad gaps in our knowledge before the creatures disappear completely.

The Dragons: Past and Present

by Daniel Cohen

Today dragons have gotten soft. Pallid, effeminate creatures like "Puff the Magic Dragon" and the Reluctant Dragon are what this once mighty tribe has been reduced to.

But in the middle ages dragons were dragons. When they were not raging about the countryside they were lurking in some dark cave guarding a stolen treasure. Occasionally a knight or saint would kill a dragon and the grateful people would erect monuments and compose poems and songs in the hero's honor.

But the dragon was more than a monster out of folklore. Two hundred years ago natural scientists regularly included the dragon in their classifications of the animals of the world. To the dragon scientists attributed characteristics scarcely less fabulous than those given by legend.

Were there ever real dragons in the world? For the answer to that question we have to go back a long way in history, back to the ancient Greeks. The Greeks had a word, *drakon,* which meant sharp-eyed or terrible-eyed. It was a name often applied to snakes whose beady-eyed glances can still be rather terrifying. The word was picked up in Latin and modified to *draco,* and from this it came into our language as dragon. To the Romans a *draco* or dragon was a very specific creature.

The first century Roman naturalist Pliny said the dragon was a giant serpent from India. Pliny wrote that a dragon of India was "so enormous a size as easily to envelop the elephant with its folds and encircle them with coils. The contest is equally fatal to both; the elphant, vanquished, falls to the earth and by its weight crushes the dragon which is entwined about it."

Allowing for exaggeration, we can be quite certain that Pliny was describing the Indian python, which grows to about 30 feet and has a strong claim to being the largest snake in the world. The idea of a python attacking an elephant is fanciful, for the snake is not nearly that large. But Pliny's description of how the "dragon" coils about its prey is an accurate account of how pythons attack smaller prey.

For centuries dragon simply meant giant snake. Somewhere early in man's history the snake picked up a very bad reputation. Ultimately the serpent came to be regarded as the symbol of evil. In the earliest piece of literature known, the "Epic of Gilgamesh," composed some 6,000 years ago in Mesopotamia, the snake is the villain. Gilgamesh is a hero who after many hardships manages to obtain a plant which would give him and all humanity the gift of immortality. At the last moment the plant is stolen from him by a serpent. Hence snakes became immortal and mankind was doomed to die. The idea that snakes are immortal probably comes from the snake's ability to shed its skin. To the ancients this must have looked like a magical process of rebirth.

The bad attitude toward snakes was continued in the Old Testament. Remember the snake in the Garden of Eden. With Christianity the words serpent and dragon became almost synonymous with Satan and the Anti-christ. It is Evil in the form of a dragon that is cast out of heaven by the Archangel Michael:

> "And the great dragon was cast out, that old serpent, called the Devil, and Satan which deceiveth the whole world: he was cast out into the earth, and his angels were cast out with him." (Rev. 12:9).

The identification of a dragon as a giant snake was common in all European nations. The old German word for dragon is *Lindwurm,* which means snake-worm or even snake-snake. The old Anglo-Saxon word *Wyrm* means dragon, serpent or worm. In

Bengal Tigers at play

California Sea Lions

Bull Elephant Seal

Bornean Orangutan (adult pair)

Lion-Tailed Macaque

Vultures in Africa

Indian Vulture

Ostrich pair with White Rhinos

Ostrich chick and eggs

Common Potto and infant

Spotted Hyena and pup

Electrode inserted near heart of shark to study shark behavior through the sensory system

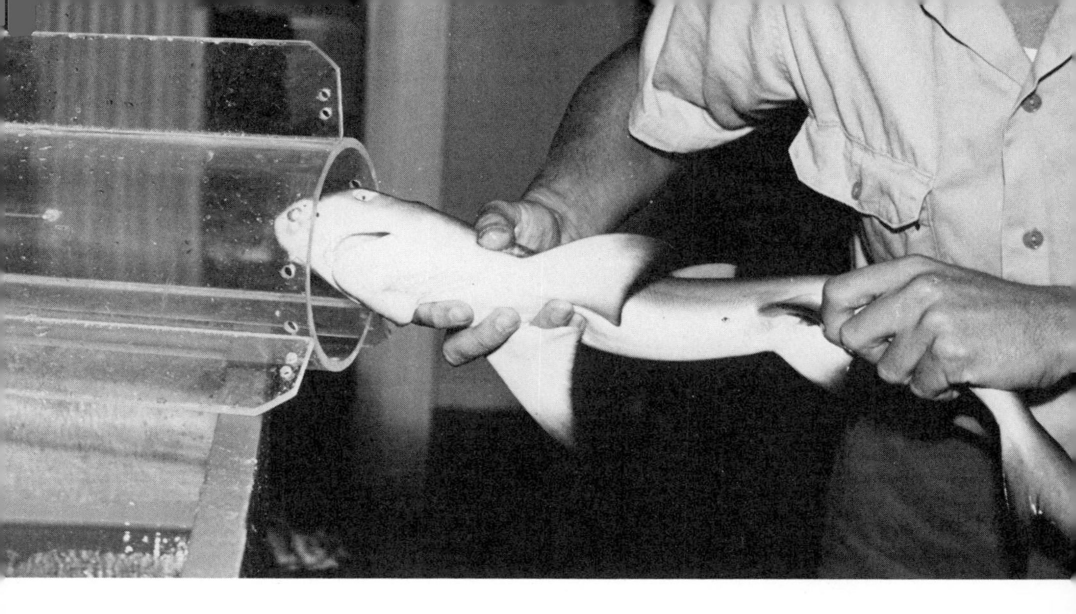

Young Lemon Sharks inserted into plexiglass restraining tubes for hearing experiments (top and bottom)

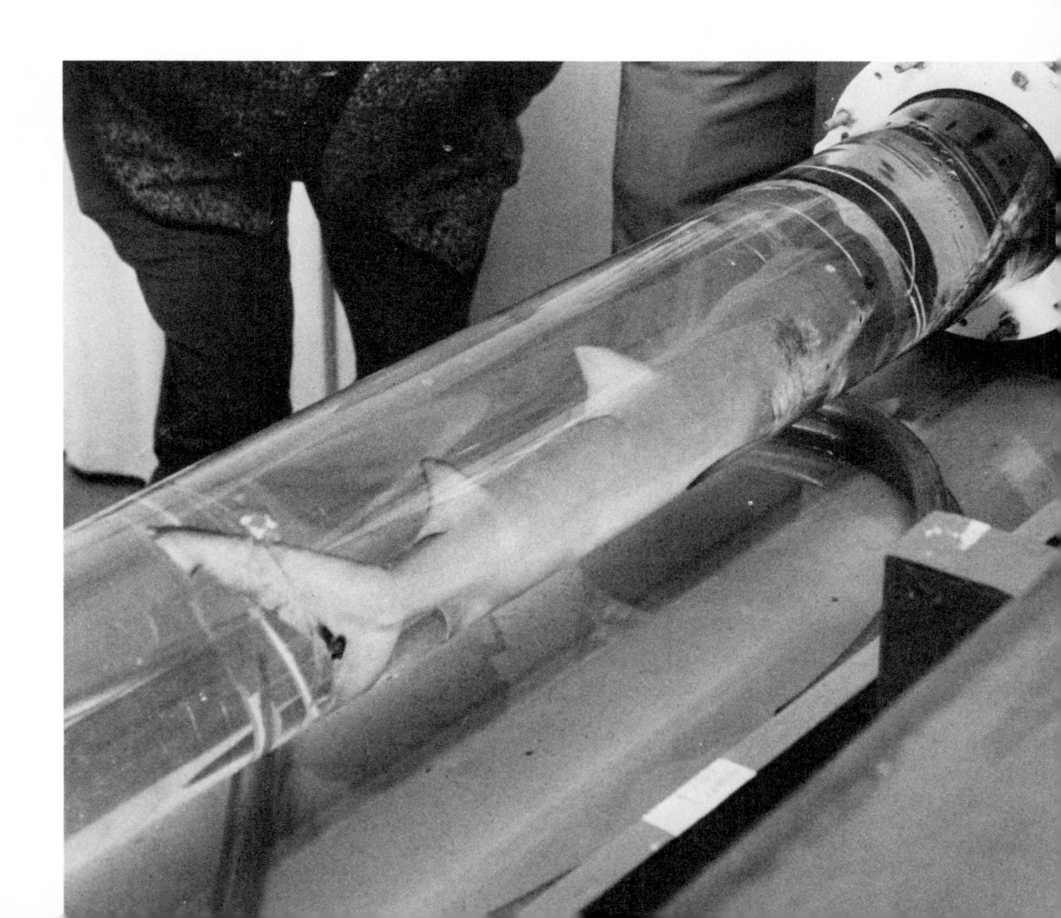

Pygmy Chimpanzee family

Striped Hyena

Adult male Lowland Gorilla

the epic "Beowulf," the hero fights a monster called "the Worm," but this is often translated as the Dragon. In old English ballads dragons are called the "laidly (loathly) Worm." In Ireland dragons were known as "direful Wurms."

In Teutonic mythology there is the famous story of the hero Siegfried (or Sigurd of the Scandinavians) who kills the dragon Fafnir. In old rock carvings Fafnir is shown as a giant snake.

So far so good, the dragon is a giant snake—where then did the dragon get its legs and wings and the ability to breathe fire and the habit of living in caves and all the rest?

The man who could answer that question is Conrad Gesner, the 16th-century Swiss physician and naturalist. Gesner compiled a multi-volumed history of the animals of the world. It was one of the most influential (and excellent) zoological books ever written and it earned for him the title of "father of zoology."

Gesner was working on Volume V, the history of serpents, when he died. His notes were later collected and published by others. An English version of his work, translated and edited by Edward Topsell, and published in 1608, was popular. It contains this description of dragons:

"They (the dragons) hide themselves in trees, covering their head and letting the other part hang downe like a rope. In those trees they watch until the Elephant comes to eate and croppe of the branches; then suddainly, before he be aware, they leape into his face and digge out his eyes. Then doe they claspe themselves about his necke, and with theyr tayles or hinder parts, beate and vexe the Elephant, untill they have made him breathlesse, for they strangle him with theyr foreparts, as they beate him with the hinder. . . . And this is the disposition of the Dragon, that he never setteth upon the Elephant but with the advantage of the place, and namely from some high tree or rock."

Despite the rather colorful language it is not hard to recognize Pliny's description of the python fighting the elephant. Naturalists of the 16th century revered the ancients, and took practically anything they had written as gospel.

After this familiar description Gesner begins talking about dragons with feet and wings. Clearly the manuscript is incomplete and the connection between the python-dragons and the winged-dragons is never made. Therein lies a mystery. To solve it we must do a little reconstruction.

Gesner quotes the Italian mathematician and physician Hieronimus Cardanus who said he saw five dried "baby dragons" in Paris in 1557. Cardanus described them thus:

"Two footed creatures with wings so small, that in my opinion, they could hardly fly with them. Their heads were so small and shaped like the heads of snakes, they were of pleasant color without feathers or hair and the largest of them was as large as a wren."

Here is our first solid mention of winged dragons. Authorities differ concerning what Cardanus had seen. Many believe the little winged creatures were manufactured monsters. There was quite a market for such curiosities. The Japanese would often sew the top half of a monkey to the bottom half of a fish and sell the result as a "mermaid." Skates and rays, flattened relatives of the shark, were often cut and embellished and sold as such mythical creatures as the "Sea Monk" and the Basilisk. Imitations like this are called, for an unknown reason, Jenny Hanivers.

It is possible that the monster makers cut off one pair of a lizard's legs and in their place sewed the wings, or part of the wings of a bat.

Another possibility is that Cardanus had seen the slightly altered and dried specimens of *Draco volans*, the flying dragon. These unique little lizards come from Java. This flying, or more properly gliding lizard has greatly elongated ribs which are connected by leathery membranes. These rib extensions can be spread apart and with the membrane stretched between them they form wings with which the creature can glide up to 15 feet. *Draco volans* was not officially "discovered" by European scientists until long after Cardanus saw his "baby dragons," but it is quite possible that some specimens reached Europe and were passed off as the young of the famous monster. *Draco volans,* by the way, has four legs like most other lizards, but these examples might have been altered to make them look more exotic.

Whatever the true origin of these "baby dragons" they were apparently very important in dragon history. At about the time they were being displayed in Paris, a French naturalist, Pierre Belon, printed a picture of a two-legged, winged dragon that looked very much like the dragon described by Cardanus. In 1640 a book called *The History of Serpents and Dragons* by Ulisse Aldrovandi was published. It contained a drawing of a two-legged winged dragon that was basically a more elegant rendering of Belon's drawing. The Aldrovandi dragon has been reprinted countless times and has become the standard for most later dragon pictures.

The origin of the dragon's reputed ability to breathe fire is easy to determine. While Gesner quite correctly stated that the Indian dragons or pythons had "little or no venom," there are plenty of venomous snakes in the world. There is no great mental leap involved in envisioning a creature with a poisonous bite to one with poisonous and ultimately fiery breath.

Gesner also mentioned that huge bones belonging to dragons were often found in caves throughout Europe. These bones, although Gesner did not know it, were the fossil remains of giant mammals—elephants, rhinoceroses and bears—that had lived in Europe 10,000 years ago.

In the city of Klangenfurt in Austria there was a monument built in 1590 which showed a naked giant in the act of slaying a winged fire-breathing dragon. (Actually in the monument, the dragon spouts water, not fire, but the intention is clear.) The monument was apparently inspired by the finding of a "dragon skull" in the vicinity of Klangenfurt. The skull was kept in the town hall and in later years scientists identified it as the skull of a woolly rhinoceros.

In 1673, a German physician, Pattrsonius Hayn, explored caves in the Little Carpathians in Hungary. He found some strange skulls which he identified as belonging to dragons. On the strength of his finds, Hayn wrote a learned paper on "Dragon Skulls in the Carpathians."

Similar finds in the caves of Transylvania inspired a German naturalist named Vollgand to write an article on "Transylvanian Dragons." Vollgand's paper was illustrated with drawings of the skulls and reconstructions of the creatures to which they were supposed to have belonged. The drawings show clearly that the bones were those of cave bears, while the reconstructions pictured the traditional winged dragon.

In the 17th century the erudite Jesuit, Father Athanasius Kircher, explained why dragons were encountered so rarely. They lived in caves beneath the earth, he said, and only rarely came to the surface. Those few individuals who were encountered above ground were poor wanderers who had somehow blundered into the sunlit world and were prevented from returning to their subterranean homes because an earthquake or some other natural disaster had blocked the passage.

When Europeans traveled to China, they found that a creature called the *Lung* was very important in the mythology of the Orient. The *Lung* looked very much like the dragon of Europe, and so the word was translated as dragon. As a result, many people began to consider the dragon as a worldwide phenomenon. In fact, the Chinese dragon is a very different sort of creature from the western dragon. In Europe the dragon is evil, in China it is good.

Although the two creatures resemble one another physically, the inspiration for the Chinese dragon was not the python or any other large snake. The Chinese dragon is always closely associated with water, and the most probable animal about which the myths were constructed is the Chinese alligator. The creature is slightly smaller than the American alligator and is quite rare now. But long ago, when the myth building began, the Chinese alligator was widespread in eastern China.

The dragon myth has been so powerful that many people have wondered if there really weren't something more to it. They could not believe that all the stories had been inspired by something as prosaic as a misunderstanding of the Latin word for big snake. These enthusiasts have put forward two candidates for the "real" dragon: a surviving dinosaur or some sort of giant lizard like the Komodo dragon.

The dinosaur idea is quite attractive because some dinosaurs did look rather like later conceptions of dragons, and because we have grown up with stories like Arthur Conan Doyle's *Lost World,* which pictures dinosaurs surviving on some unknown island. Unfortunately, all evidence indicates that dinosaurs died out 70 million years ago without any descendants.

The theory that mankind has somehow retained the "concept of the dinosaur," which was reflected in later dragon myths, by "racial memory" is just plain silly. There were no men 70 million years ago.

The Komodo dragon is a powerful 12-foot lizard that is found in a remote part of the East Indies. It might well inspire dragon legends. It was, after all, called a dragon by those western explorers who first encountered it. But the Komodo dragon was not "discovered" until 1912. This does not mean that garbled accounts of its existence, or even preserved specimens of the creature itself did not reach Europe long before this century. But there is not a scrap of evidence that they did.

Besides, such theories are really unnecessary. For centuries everyone knew that the dragon was a large snake. Only in comparatively recent times did the dragon get legs, etc.

As far as history is concerned it looks as though the dragon might have the last laugh. The Roman Catholic Church announced that after careful historical study there was considerable doubt as to whether a number of popular saints ever really existed. Among the doubtful was St. George, the most famous dragon slayer of all times. For a long time people have spoken of St. George slaying a mythical dragon. Now it may turn out the dragon is more real than St. George.

Spotted Hyena: Deadlier than Lions or Tigers

by George W. and Lory Herbison Frame

Recently a six-year-old South African girl lay sleeping on the verandah of her home because Malawi's summer heat made it intolerable indoors. Silently and without warning a spotted hyena crept out of the nearby bush, entered the verandah and with one bite crunched the life from the child. Within seconds, other hyenas rushed in on the inert body. Although the girl's parents had heard the noise and beat off the attackers, they were far too late.

Although man-eating spotted hyenas *(Crocuta crocuta)* are not common, in some areas they kill more people than lions and leopards combined. In this century they have mutilated the faces or severed the limbs of hundreds of Africans, and a few Europeans as well. Others died after severe infection set in from the bites alone. One African, however, lived a reasonably healthy life after a hyena bit off his nose, palate, upper teeth, tongue, and most of the lower jaw. He had slept outside his hut on a hot night and the hyena took just one bite with its powerful jaws.

Oddly enough, humans are largely responsible for creating man-eating tastes in hyenas. In 1898–99, according to A. B. Percival in *A Game Ranger's Notebook,* a smallpox epidemic in Kenya killed thousands of people. Dead bodies, as well as old and sick persons near death, were abandoned in the bush specifically for hyenas and other scavengers to devour. Once the hyenas had learned that sick men were easy prey, they became very bold. When the famine and epidemic abated, the number of human bodies were insufficient to feed the growing numbers of hyenas, so they began attacking sleeping adults and children. This is just one of the many examples that have been repeated over the years.

Man's attitude toward the hyena has been a mixture of reverence and loathing, clouded until recently by a lack of real understanding. Witch doctors often kept them as pets, and some trained their hyenas to eat men or rented them to persons seeking revenge. Before Tanganyika became Tanzania (1964) Game Ranger George Rushby reported shooting hyenas that had beads interwoven in their hair and mysterious patterns cut into their skin; all the work of witch doctors. One hyena he shot was claimed by a woman witch doctor as her lover.

Hyenas, contrary to popular belief, are not dogs or even related to dogs. They're actually relatives of civets and mongooses.

In modern Zululand hyenas are in great demand for magic medicines. Most body parts are used in one way or another, and special significance is attributed to fats, eyebrows and paste from the anal gland. Hair and skin are burned, then ground into powder and swallowed for stomach complaints.

There's valid reason for witch-doctors to adopt the hyena as a tool of their trade. The eerie, spine-tingling "0000-whup" of its whooping call in the night is enough to raise the hair on anyone's neck. Its wolfish appearance and spectral eyes, on a moonlit night, are shatteringly impressive.

The spotted hyena is bigger than either the striped hyena *(Hyaena hyaena)* of northern Africa and Asia or the brown hyena *(Hyaena brunnea)* of southern Africa. Its fur is ochre yellow with circular dark spots over much of the body. From the massive head, neck and forequarters the body slopes backward to smaller hind-quarters. Its movements are ungainly. Adults grow about five feet long, with a shoulder height of 30 inches or more. Three adult spotted hyenas in Tanzania's Ngorongoro Crater weighed about 110 pounds each, but in other areas they are reported to reach 190 pounds.

A common African belief is that spotted hyenas are hermaphroditic and can change sex. This fallacy probably results because the female external sex organs are unusually extended and developed so as to resemble a male's. Mature females, however, have two large nipples on the abdomen to distinguish them.

In the 1960's, the first objective scientific research on hyena behavior and ecology was begun in Africa. Dr. Hans Kruuk, a Dutch scientist, was the first to study the social organization of spotted hyenas. He learned they have a matriarchal society made up of clans. In the 110 square miles of Ngorongoro Crater he found 510 hyenas living in eight clans, each consisting of five or six families.

As with many mammals, scent-marking is of great social significance in establishing territorial rights between clans. The spotted hyena deposits scent by crouching its hind quarters and slowly moving forward as a stem of grass rubs between its hind legs. Scent glands protrude in a bulge, just above the anus, as a smear of strong-smelling secretion is left on the stem to serve as an olfactory "signpost." Generally other clan members mark the same spot, the dominant individuals first. As they leave, lower ranking hyenas take their turn. Each clan, led by a female, regularly patrols and scent-marks its territorial boundaries.

Jane and Hugo van Lawick, who continued Dr. Kruuk's work in Ngorongoro Crater, found that boundaries of clan territories are not permanent but gradually shift. One clan doubled its territory in a year by fighting the neighboring clan. The seriousness of border intrusions was demonstrated once when two spotted hyenas trespassed only a few yards into the adjacent territory and lay down to sleep. The resident clan discovered the intruders and attacked viciously. One hyena escaped across the border into his territory, but his slower companion was bitten badly and probably died. So strong is the territorial drive that three residents of an area can chase away 20 or more intruders.

When a wildebeest or other prey is killed on a clan boundary, the disputed meat sometimes changes clans several times before all is devoured. If a clan chases its prey across a boundary and makes a kill outside its territory, the resident clan will attack and claim the meat.

When two spotted hyenas of the same clan approach each other they hold their tails between their legs and pull back the corners of their lips in a grin. The dominant animal usually responds with a wide yawn. They then push under the chin, rub against the chest, and sniff each other.

Cubs are generally free of a clan's territorial restrictions until 12 to 18 months of age. As they mature, participation in marking parties and border skirmishes shapes their territorial behavior. Sometimes a male may become a member of two clans, holding high rank in the clan of his birth and a lower rank in the neighboring clan.

Spotted hyenas obtain much of their food by killing their own prey, exploding the myth that they are cowardly scavengers. Often they hunt in packs at night or on cloudy days, but sometimes they hunt singly or in twos or threes. Larger packs have better hunting success and generally select larger prey. For example, Dr. Kruuk saw 21 lone chases, only four of them successful, whereas nearly all pack hunts ended with a kill.

Spotted hyenas have an uncanny ability to detect the slightest abnormality or weakness in an animal's behavior, and individuals that appear normal to the human eye are recognized by the hyena as easy prey.

A pack of spotted hyenas on a hunt doesn't necessarily chase the first animal encountered. Often they ignore abundant prey and lope for miles, stopping frequently to scentmark the grass. When the pack arrives at the chosen hunting area, the hyenas stop and look around. They select their prey, then creep forward as closely as possible before attacking.

Near Ngorongoro Crater's Mandusi Swamp we followed a pack as they chased an adult wildebeest. Two hyenas snapped at its flanks as they pursued nearly a mile along the swamp's edge. Suddenly the gnu leaped down a vertical bank into a stream. The leading hyenas followed it into the water, tore open its flanks as it floundered, then pulled it ashore. In seconds hyenas covered the carcass. Latecomers pushed their way in among snapping mouths. Within 30 minutes the 400-pound gnu was reduced to stomach contents, hooves and skull.

During the wildebeest calving season many calves are caught a few minutes after birth. Food is so easily obtained then that hyenas hunt during the day as well as at night.

On the Serengeti Plains and within Ngorongoro Crater the wildebeest makes up two-thirds of the hyena diet. The remaining third is primarily zebra, Thomson's gazelle and Grant's gazelle. Each adult hyena eats an average of 4½ to 6½ pounds of meat a day.

Sometimes hyenas kill an animal but lose it to larger predators such as lions. In Ngorongoro Crater during a single night Dr. Kruuk saw the prey change ownership several times, with the determined hyenas snapping at the feeding lions.

One of the spotted hyena's special rules is to respect scent marks of clan members. If a pack kills more than it can immediately eat, it puts scented feces around the carcass. Other packs and solitary hyenas of the same clan observe these marks and do not eat the meat, regardless of their hunger.

Although spotted hyenas are predators, they also are superbly equipped for scavenging. Keen hearing enables them to determine direction accurately, and their sharp eyes can spot distant vultures gathering around carrion. Their jaws are so powerful they can crack the toughest marrow bones—even those of elephant and rhinoceros—and their stomachs seem capable of digesting almost anything. Hyenas steal shoes, pots, garbage, even canned foods, and they are notorious for digging up human graves.

A spotted hyena killed in the Kruger National Park in the Transvaal some years ago had in its stomach the tail of an impala, feet of a leopard tortoise, leaves, green peas out of a can, chunks of stiff corn meal porridge, scraps of paper, hair, bone fragments, seeds from a wild fruit, fragments of a car tire and the rubber nipple of a baby feeding bottle!

At Ngorongoro Crater Jane and Hugo van Lawick found that a spotted hyena clan's social life focuses on dens with resident cubs. Dens often are abandoned termite mounds or animal burrows, and sometimes are shared with a porcupine, wart hog or bird. The female lies outside her den almost all day, and mothers with small cubs normally keep adult males at a distance, probably because of occasional cannibalism.

Unlike most carnivores, spotted hyenas are in an advanced stage of development at birth. Their eyes are already open and many teeth are visible. They are active and use their front paws to pull themselves along. At first their mother enters the den to suckle them, but by the tenth day they come to the den entrance to suckle.

An 18-month nursing period is essential because the female spotted hyena does not lead her cubs to kills and does not regularly bring food to the den as do some carnivores such as the Cape hunting dog. Cubs must depend on a steady diet of milk. But female hyenas in the Serengeti must leave their suckling cubs for several days at a time while they seek food, and the cubs often are emaciated and sometimes dead when the mother returns.

The cubs are notorious for their temper tantrums during the weaning process and may squeal and dash about in anger and frustration for several weeks. During a weaning tantrum one cub ran around its mother holding its tail erect, grinning and squealing harshly. When the cub tried to nurse, its mother gave it a quick nip. But the youngster persisted for almost an hour before the harassed mother seemed to lose all patience, and followed her cub in tiny circles, repeatedly biting its neck and back.

Disciplined research has provided a comprehensive picture of the spotted hyena, and an appreciation of it as a predator and scavenger. Before 1950 it was destroyed as vermin over much of Africa. Today these animals are protected in many national parks and wildlife sanctuaries as an important part of the ecosystem.

Those Bloodthirsty Vultures

by George W. Frame

With legs outstretched and tail held down to slow its descent, the White-headed Vulture soared down through the early-morning sky. It hit the ground rather heavily, a few yards from a dead wildebeest. Then with awkward lopes, rocking from side to side, it sprinted toward the carcass.

Apparently the wildebeest died from disease or just plain old age, for it had no external injuries. As the vulture inspected the tough unbroken skin, it seemed unsure where to begin. It pecked out one of the eyes, and then waddled around to the other end of the carcass. Here the skin was softer and the bird had little difficulty tearing into it with its sharp beak. As the hole deepened, the White-headed Vulture literally plunged its head into the bloody flesh and tore pieces away.

Within minutes, dozens more vultures were circling as tiny dots high in the sky. Quickly they plummeted earthward, bouncing in sodden-like heaps here and there around the carrion. As I watched, the carcass became completely buried under a thick undulating blanket of feathers.

Amid much hissing, squawking, squealing and ruffling of wings, more than 80 vultures fought for their share of the meat. Another half as many looked on from the sidelines. Sometimes two vultures pecked viciously at one another and jumped into the air to clash talons. But one was always quick to retreat, and I never saw one injure another.

In Tanzania's Serengeti National Park as many as six different species of vulture feed together. These are the White-backed Vulture *(Pseudogyps africanus)*, Rüppell's Griffon *(Gyps rüppellii)*, Lappet-faced Vulture *(Torgos tracheliotus)*, White-headed Vulture *(Trigonoceps occipitalis)*, Hooded Vulture *(Necrosyrtes monachus)* and Egyptian Vulture *(Neophron percnopterus)*. Rarely are they seen eating anything but carrion, which means that all six of these species compete for the same food resource. They compete keenly, and are able to eat an entire wildebeest or zebra in less than an hour.

Dr. Hans Kruuk, a Dutch zoologist, recently studied vulture feeding behavior in the Serengeti National Park. He found that both the Lappet-faced and White-headed Vultures most often feed by tearing at the tougher parts of the carcass; they hold with their claws and tear with their beaks. The White-backed Vulture and Rüppell's Griffon, by contrast, mainly pull out the soft fleshy parts of the carcass. And the Hooded and Egyptian Vultures both peck scraps from the ground, almost like chickens. Thus, Dr. Kruuk found that the six vulture species could be classified into three feeding groups, each specializing on a certain type of food from the carcass. This classification of food habits is also strongly substantiated by the shape of the skull, beak and tongue, and the size of the vulture's body.

There is some evidence from a number of observers that Lappet-faced and White-headed Vultures sometimes kill their own prey. Dr. Kruuk, for example, twice saw a Lappet-faced Vulture feeding on a dead but still warm young Thomson's gazelle with no other predators nearby. He also found two White-headed Vultures with a very freshly killed bat-eared fox.

The White-backed Vulture and Rüppell's Griffon rarely touch any food other than carrion. These are the two species which specialize in pulling out soft meat from carcasses. Both have a very long, almost naked neck, which is well-adapted for reaching far inside a large corpse. When soft parts are too deep to reach, these determined scavengers crawl right into the carcass through any small openings they can find.

Hooded Vultures are occasionally predators, but on very small prey. W. Fisher, a German zoologist, reported that Hooded Vultures sometimes catch lizards and rats and that they also eat termites, grasshoppers and other insects. From Zambia comes a report suspecting Hooded Vultures of eating eggs in a flamingo colony. Otherwise, they are commonly seen around native villages and the garbage dumps of cities, where they feed extensively on carrion and garbage.

Recently Egyptian Vultures were brought to world-wide fame when Jane and Hugo van Lawick photographed one tossing rocks from its beak to break open ostrich eggs. A South African newspaper reported exactly the same thing in 1868, but it was forgotten by science for an entire century. The Egyptian Vulture also catches termites, grasshoppers and dungbeetles. It is the only vulture which Dr. Kruuk often saw feeding around hyena dens—picking at bones and hyena regurgitations. In some areas, Egyptian Vultures are closely associated with human settlements, feeding on excrement and garbage.

The arrival of the different species of vultures at carrion follows a definite pattern. Dr. Kruuk found that in 50 percent of his observations on the Serengeti plains the White-headed Vulture arrived first, although this species represented only three percent of the vulture population. At the time when most vultures are present on the carrion, the White-backed Vultures were most numerous. When nearly all the vultures had finished and departed, the Hooded Vultures were still cleaning up the last scraps.

Why certain vulture species arrive first on a kill is not certain. Their arrival time may be related to feeding habits on the kill; or it may be related to searching behavior. Vultures have been seen at heights up to nearly 12,000 feet, and it is likely that different species fly at different levels. Dr. Kruuk is certain that at least some of the differences in arrival time result from some species finding the carrion themselves (White-headed and

Lappet-faced Vultures), and other species (White-backed Vulture and Rüppell's Griffon) being attracted by other vultures landing.

Several years ago, Dr. Colin J. Pennycuick used a glider plane to study vulture soaring behavior over the Serengeti plains. During the morning and early afternoon he followed the vultures as they patrolled over the herds of migratory ungulates (hoofed mammals) in search of food. The searching vultures generally flew slowly (about 60 feet per second or less) with their heads pointing downwards, looking from side to side. When the migratory herds moved far from the vultures' roosts, the vultures often traveled 45 miles or more, one way, to feed. They made such trips in about 96 minutes, moving from thermal to thermal, generally staying less than half a mile above the ground. By soaring in thermal air currents, vultures expend very little energy.

Little is known of which sense organs vultures use in locating their food, except that vision is obviously most important. There seems to be no evidence that smell is of any significance to African vultures, as it is to the American Turkey Vulture *(Cathartes aura).* Dr. Kruuk made one observation which shows that Hooded Vultures are able to react to predator noises: He used a loudspeaker to play tape-recorded sounds of hyenas feeding on a kill. The result was that Hooded Vultures gathered around his Land Rover even before the hyenas arrived.

Most fights among vultures feeding on carrion appear to be between members of the same species. When vultures fight they strike at each other with their bill, and often they use their powerful claws by jumping into the air and throwing their legs forward. Wings are usually held open during fights, but are not used as weapons.

The Lappet-faced Vulture shows the most conspicuous threat postures. As it very slowly approaches other feeding vultures from a few yards away, it walks upright, bill pointing down and wings half-opened. Or sometimes it approaches in a horizontal posture with wings wide-open, tail held upwards and neck feathers raised. Occasionally it stretches its very long neck forward and upward, and utters a loud hiss.

Dr. Kruuk observed 135 aggressive encounters where a vulture of one species attacked a vulture of a different species, and in all but two instances the attacker won. Consequently, aggressive species such as the Lappet-faced Vulture and Rüppell's Griffon are very successful in stealing food from other, less aggressive, vulture species.

Large mammalian predators such as spotted hyenas, lions and cheetahs often snap at vultures if they try to feed on their kill. The White-backed Vulture and Rüppell's Griffon seem braver than the other vultures in this respect, for they try to feed off the carcass in spite of the danger.

Although vultures often benefit from the kills of predators, the reverse is sometimes true. Hyenas and jackals frequently respond to the sight of vultures landing by running towards them. Sometimes they find the vultures feeding, but other times they discover that the vultures have landed to rest or to drink. More than once, Dr. Kruuk has seen a sleeping hyena come alert at the sound of a vulture landing. Wind swooshed through the bird's feathers in a loud and distinctive manner; the hyena jumped up, looked in the vulture's direction, and ran toward it in anticipation of food.

Vultures are to many people the most repulsive of birds. They are considered dirty because they feed on carrion. They are unappreciated because they are not very pretty. But vultures are a vital part of many ecosystems. They perform important functions such as cleaning up carrion and helping to recycle nature's nutrients as quickly as possible.

There seems to be only one reliable record of a vulture killing a man. The unfortunate victim was the Greek poet Aeschylus, who, in 456 B.C. (at the age of 69 years) was killed instantly when a Lammergeyer, or Bearded Vulture *(Gypaëtus barbatus)*, dropped a tortoise on his head. The Lammergeyer was probably trying to break the tortoise shell so it could eat the contents. A very unusual record for an unusual kind of bird.

A Watched Potto Never Grows

by Myrdene Anderson

What's a potto? Pottos are small, arboreal, nocturnal primates which ordinarily live in the forest canopies of tropical Africa. The scientific name is *Perodicticus potto,* the species designation adapted from their name in Twi, a language of the Gold Coast. In Dahomey, the word for potto could be translated as "softly, softly," apt for an animal that moves inconspicuously, vocalizes infrequently, and even looks soft and muffled in its reddish-brown fur.

The closest relatives to pottos are the loris and galago lemurs found in south Asia and Africa, respectively. These animals differ from the lemurs of Madagascar in social habits and in the specialization for their ecological niche. Pottos have been described as slow, deliberate climbers and solitary. But the habits of the potto family described here call that picture into question.

The two pottos I studied, Piglet, the female, and Eeyore, the male, were approximately the same size, each weighing under two pounds and measuring 11 inches from crown to rump. Their bodies were covered with close fur, reddish brown on the back and gradually changing to grey on their bellies. Each had a short, bottle-brush tail about two inches long and a pair of long, slender arms and legs. The index finger-toe on both the hands and feet is nothing more than vestigial stump and on the foot it is equipped with a long grooming claw. Another distinctive feature of the potto are the several ⅛-inch protrusions, called cervical spines, found along the back of the animal's neck.

When Piglet and Eeyore were transferred to my home in June 1970 it appeared that Piglet was pregnant and, since I knew neither the date of conception nor the gestation period for pottos, imminent birth was possible.

For a few days after my pottos arrived I watched the mutual adjustment made by them and my cats. Neither set of animals showed animosity and within weeks the potto cage was enlarged to become a potto room. Finally, potto territory and cat territory were one and the same.

At first the two arrivals lived on their familiar fruit, live crickets and newborn mice which were put out in early evening.

Anything extra sweet, such as honey, was a treat. Delicacies were used to coax the pair awake for their numerous daytime visitors. Piglet especially loved to raid the kitchen at night and developed a new range of favorite foods. She seemed to crave cat food, fig newtons and marshmallows, and would try nearly anything not carefully wrapped.

Just as she had done in the cage, Piglet continued to descend to the floor of the house. Her mate, Eeyore, was not to do this until September. When going from a height to horizontal surfaces, the pottos did move rapidly. Locomotion on the floor was always quadrupedal, the legs extending the rump higher than the shoulders, giving a bearlike waddle to their gait.

Both Eeyore and Piglet enjoyed contact with other creatures, man and animal alike, and seemed to like attention especially when it involved grooming.

Friends who came to visit and observe the pottos were skeptical of the renewed evidence of pregnancy that manifested itself in October. After the apparent false alarm in June, most of us gave up looking forward to a newborn potto. What we did not know was that the usual gestation period is about six months.

Finally on the evening of November 8 I came home to find a bulging Piglet nestled on a shelf of clothing. She accepted tidbits of food but would not stand or come out. The next day I was absent from the house for no more than three hours and I returned home to find a newborn potto, whom I named Tigger, crawling into the world.

He weighed just over an ounce and measured approximately 3¼ inches from crown to rump. Tigger's eyes were large and alert from birth. His nose was a pink button and his fur, cream colored at birth, would take on the shadings of the adult potto.

Tigger was always attached to Piglet's front and forcibly removing him was difficult. He could not be taken from his mother for exact measurement for several weeks. To add to this difficulty, measurement was highly arbitrary. Tigger would grow an inch and retract two. There were similar problems with his weight, partly due to an erratic postage scale.

Tigger gradually developed patterns of behavior typical of all pottos and any trouble-prone newborn: the concern over grooming, acrobatic deftness in hanging from branches or climbing a curtain rod, butting and teasing its mother, lunging at a cat, and, on the two-month anniversary of its birth, eating his first solid food, a marshmallow.

After the first few weeks Tigger's growth levelled off abruptly, most likely because Piglet's lactation during this period decreased. In spite of this slowdown in growth Tigger's health appeared excellent and there seemed no real cause for concern.

One day, however, Tigger was found in critical condition lying on the floor, too weak to hold to his anxious mother. Forced feeding of warm milk revived the infant and this was continued to supplement the mother's diminishing supply.

Tigger never really took to formula feeding and he resisted any nipple substitute. The only thing he seemed to like about feeding time was the opportunity to climb potto-fashion from hand to hand and finger to finger.

However mature Tigger was mentally, physically he was still a dwarf. His weight after 70 days was not yet three ounces, only a quarter of that reported in growth curves for other pottos. Such a discrepancy was serious although it did not seem to cause a great deal of concern to Tigger.

Besides his low weight gain, Tigger may very well have been a runt at birth. His parents were slighter than most pottos, weighing roughly 25 percent less than reported averages.

So, although Tigger prospered, it was not on a grand scale. At 10 weeks he looked very much like a furry frog. Behaviorally he is well on his way to becoming an adult potto, or at least a potto pigmy.

Postcript. At four months Tigger weighs in at 10 ounces. Indications now strongly suggest that he is a girl. Such reconsiderations are not uncommon with adult pottos.

Are Chimps Really Animals?

by Daniel Cohen

What differentiates man from his primate ancestors, according to scientists, is that man can make and use tools. Now, as a result of the careful observations of a young Englishwoman, the great anthropologist, Dr. Louis S. B. Leakey, has said, "We must either redefine man—redefine tools—or accept chimpanzees as men."

The young woman is Jane Goodall, who at the age of 26 and without any special training, began what Dr. Leakey has called "the most remarkable study on any primate ever made."

It's not just that her discoveries will help us understand the behavior of chimpanzees better, and our own behavior as well. Jane's five-year expedition also contains all the elements of a roaring good story: frustration and triumph, heartbreak and humor, danger, high adventure and that rarest of all elements—genuine romance.

In July, 1960, Jane Goodall, with Dr. Leakey's encouragement, began her lonely task of observing the daily lives of the chimpanzees in Tanzania's rugged Gombe Stream Reserve. Chimpanzees, those animals considered closest to man, had often been studied in zoos, but because in the jungle they are extremely wary of man, powerful and potentially dangerous, their life in the wild was practically unknown.

Winning the confidence of the chimps so that she could get close enough to observe them took enormous patience. Dr. Leakey uses Jane's success as sort of an inspirational story for other scientists. He told a meeting of the world's leading anthropologists, "You must have a dedication and a willingness to go all out through illness and bad climates, always with the knowledge that despite your dedication, you may not get the answers at all. On this matter of devotion, let me remind you that Jane Goodall is now in her fifth year of studying the chimpanzees, and in the first eighteen months she got none of the important results she has obtained since."

What are the important results? Says Jane, "The fact that these chimps use twigs and grasses when feeding on termites was one of the most exciting discoveries I made. It was known that some wild animals use natural objects as tools—but the chimpanzee, when he strips leaves from the twig, is actually modifying a natural object to suit it to a specific purpose . . . and he is thus making a tool."

Chimps also crumple leaves to make "drinking sponges." These home-made "sponges" are dipped into water and the water that clings to them is sucked off by the chimp. The crumpling of the leaf, Jane notes, increases the absorption. It is another example of how these highly intelligent primates are able to adapt natural objects to a specific purpose, another example of tool-making. They also use leaves as napkins to wipe sticky food off their hands and mud off their feet.

Another important discovery she made is that chimpanzees regularly catch and eat other animals, usually small monkeys, but sometimes animals as large as young deer.

Jane Goodall's observations have a direct bearing on a school of thought that has been developing in recent years among students of man's evolution. Evidence has accumulated that man rose from a line of carnivorous apes, and that he first used altered objects as weapons not tools. In addition, some studies of primate behavior seemed to indicate that many of the higher primates were unusually aggressive—even vicious—types of animals. Thus, the thinking ran, man may be "a natural killer" and this killer heredity may explain mankinds seemingly irrational attachment to wars and other forms of violence.

Jane's discovery that chimps are occasional hunters seemed to support the natural killer theory. But writing in *National Geographic* magazine, she says, "In some scientific circles a controversy turns on the question of whether early man first used objects as tools or as weapons. One certainly cannot draw concrete conclusions from this chimpanzee community. But the examples I have given amply demonstrate that these chimps, though seldom using objects as weapons, have reached a high level of development in selecting and manipulating objects for use as tools."

In general, observations at the Gombe Stream Reserve indicate that wild chimpanzees are very much as we would like to think they are—intelligent, individualistic, cooperative, affectionate and, even as adults, playful.

There is, of course, a definite social hierarchy among the chimps, but it is not a brutal oppressive one. Real fighting among chimps is rare, even during the mating season. Jane describes a classic meeting between two giant male chimps, "Catching sight of each other, the two friends ran together and stood upright face to face, all their hair on end. They looked magnificent as they swaggered slightly from foot to foot before flinging their arms around each other with small screams of pleasure and excitement." Aggression is worked off in ferocious but harmless displays, rather than real combat. The chimps have even worked out a fairly tolerable state of peaceful coexistance with their nearest relatives and competitors, the baboons.

Jane Goodall initially intended to start her study accompanied only by an African cook and his family. Tanzania authorities, however, insisted that she bring along someone else because the jungle was a dangerous place for such a young girl. So she brought her mother. The authorities soon became convinced of her competence and seriousness and her mother returned to

England. Home during the difficult early months of the study was a simple tent, and at that time the question of whether or not finances for a lengthy study would be available was even in doubt. After three preliminary months, the National Geographic Society took over sponsorship of the research and financed it for another 20 months. Those 20 months stretched out to five years. In 1965, work had been completed on three semipermanent buildings in the area, where research will continue on a long-term basis. Jane feels that at least 10 more years of observation are necessary to trace the complete life history of the chimps at the Gombe Stream.

Help came from other sources. Enda Koning, a Dutch girl living in Peru, had been so inspired by Jane's first article in *National Geographic* that she scraped together her own fare to Africa so she could be part of the project. Later, they were joined by Sonia Ivy, an assistant secretary, to help with the monumental amount of paperwork.

How about the romance? In 1962, Dr. Leakey suggested to the Dutch nobleman and noted wild-animal photographer, Baron Hugo van Lawick, that he should film Jane Goodall's chimpanzee friends. The Baron's photographs are not only some of the most remarkable wild-animal pictures ever taken, they also form an invaluable part of the scientific record. His color motion pictures were the basis of a CBS-TV special on Jane Goodall's work.

The Dutch nobleman and the English scientist were married on March 28, 1964, in London, where Jane, a candidate for a Ph.D at Cambridge University, had gone to finish a term. They limited their honeymoon to three days. They rushed back to Africa because they had received word that one of the chimps that they had been observing for a long time had just had a baby. Romance or no, they wanted to chart the development of an infant chimp in the wild from the very beginning.

Dr. Hibben's New Mexican Ark

by Robert Gannon

Some strange-looking animals are roaming the wastelands of New Mexico nowadays. In the isolated, arid Florida mountains in the state's southwest corner, goat-like Siberian ibex clamber over the rugged rocks. Sixty miles west, near the Arizona border, kudus and gemsbok contentedly munch mesquite and creosote bush as their numbers gradually increase. And in the opposite corner of the state, 450 miles northeast, some 1,200 North African Barbary sheep ramble through the huge Canadian River Canyon. None of these animals really belong in the U.S. All have been imported. But if their destruction continues in their native Africa and Asia, the American Southwest may be the only place in the world where future naturalists can see them in the wild.

The animals are there because of New Mexico's "Safe from Extinction" program, and the project's success is due to one man: Dr. Frank Hibben, 57, a professor of anthropology at the University of New Mexico. He's an adviser to the Federal government on African affairs, he's written half-a-dozen books on archaelogy, and he's the director of the university's Museum of Anthropology. But his real passion is the saving of some of the world's threatened animal species.

"Some of the most beautiful and spectacular animals of the world are hovering on the verge of extinction," he says, "while here in New Mexico we've got thousands of arid, rugged, *empty*, square miles where some of these large animals can thrive. Let's put this land to use.

"We have to hurry. if we don't act fast, nothing will be left but stuffed heads on the wall to show us what magnificent beasts they were."

One animal not wholly out of danger of extinction is the Barbary sheep, and early this year Dr. Hibben took me into the Canadian River Canyon to see some of them. We searched the gorge all day, scanning the rugged cliffs. Then, as dusk began to press, Dr. Hibben suddenly shouted "Barbarys!" He whipped his field glasses from their pouch, leaned his elbows on a rock and peered upwards. Four hundred feet up the cliff five shapes moved in the dusk. They stopped—and disappeared, blending into the background. Suddenly one old buck flexed his legs and elevated himself five feet, straight up onto a boulder. He stood there, silhouetted against the sky for a moment. Then he shook his great curved horns, turned and all five animals melted off, disappearing into the rocks.

"Beautiful," Hibben murmured. "Beautiful." He replaced his glasses. "These were the first animals we concentrated on, you know, and frankly, the whole Barbary project was an experiment. We could have completely failed. The sheep could have simply disappeared, or multiplied out of control. But it worked; the experiment succeeded."

It was a success on three counts: (1) The foreign sheep adapted to the American Southwest. (2) If they do become extinct in Africa, the species will be saved. And (3) because of the tourists the sheep are bringing in, the state is profiting economically.

The idea of saving threatened animals first came to Hibben during World War II, when as a Naval officer he spent time in Algeria, near the Barbary sheep's home. The species was nearly extinct because of the war and massive slaughter due to lack of other fresh meat in the area.

"Those rugged, craggy parts of the Atlas mountains had reminded me of sections of New Mexico, and it occurred to me that if I could only get hold of some Barbarys and turn them loose here, maybe they could flourish. And then I thought, why limit it to sheep? Why not try to save other species threatened with extinction.

"I visualized a herd of gemsbok—huge, purplish African antelopes—galloping across the terrain; and thought, wouldn't that be something?"

He began selling his grand idea to other conservationists and sportsmen. "My major problem at first was with the state's fish and game people. They couldn't conceive of the idea. They said, 'We've got 'em in zoos; what are you worried about?' I told them that I wanted to turn them loose as wild animals. Creatures in zoos are under very unusual conditions, you know, because of confined space, unnatural food or tensions, and they often won't breed. 'I want them running free,' I said."

Gradually, his enthusiasm caught hold. Then, in 1964, when he became president of the New Mexico Wildlife Conservation Association and Chairman of the State Game Commission, he was in a position to develop plans.

The Canadian River Gorge, he decided, was an ideal location for a herd of Barbarys. A 60-mile slash in the rolling plain, the canyon drops a thousand feet to the frothing river, and from the main gorge hundreds of miles of tributary canyons jag off through the sand. Only scattered, stunted pinion pine and staghorn cactus stud the red and cream sandstone chasm.

A few deer, antelope and adventurous cattle wander along the steep trails, but they can't navigate the canyon's sides. Barbary sheep can.

Still, not everyone agreed with Hibben that the sheep should be turned loose. "Whenever I told someone of the idea he'd think of the starling and English sparrow and rabbit of Australia—all imports, and all pests—and I'd hear discouraging words."

But one by one Hibben convinced the others on the game commission. Finally, they voted money to buy sheep from a number of American zoos. It wasn't enough. So the anthropologist became a fund-raiser, gathering most of the rest from such conservation-and-hunting organizations as the international Shikar-Safari Club. Still there wasn't quite enough. So Hibben dug into his own pocket. "It really took all my wherewithal for several years. But I was so convinced that we must save these animals I had to put in all I could spare. Lots of times I'd tell my wife we couldn't afford something, and at the same time I'd be spending money to get some animal."

When the Barbarys were released in 1950, Hibben was a lot more worried than anyone else knew. If they died off or got out of hand, his whole dream would collapse. But the gamble paid off: The herd grew quickly, then stabilized at between one and two thousand.

With the success of the Barbarys, Hibben began concentrating on other threatened species, those particularly suited to fill New Mexico's "ecological niches."

The Siberian ibex was particularly adaptable to one of these niches, and Dr. Hibben immediately plunged into the complex task of obtaining eight of the animals from a remote Himalayan village. They ended up costing about $6,500 each, much more than had been estimated, but everyone was delighted anyway, so much so that Hibben was encouraged to begin a search for another rare variety of ibex—the brownish, highly-intelligent Iranian, probably the 10,000-year-old ancestor of our domestic goat. It lives only in the rugged mountains of northern Iran.

"The two things that impressed me most about Iranian ibex in the wild," says Dr. Hibben, "is their remarkable eyesight and their climbing ability. They get away from you by bounding up the roughest, most inaccessible rocks you ever saw. They go up places where you wouldn't think a fly could walk. And their eyes—they're *phenomenal.* Once, at dawn in Iran, I was hidden in a valley, and through field glasses happened to see three animals on top of a distant ridge outlined against the sky. They were all looking right at me."

Iranian ibex are also quickly disappearing, killed off by

newly-armed tribesmen for their fine, silky undercoating of cashmere. Back in the mountains, a few herds hang on.

But how could Hibben, teaching classes in Albuquerque, arrange for their capture? The answer came in a letter from Major William T. Moore, an Air Force man stationed in Kermanshah, in western Iran, and one of Hibben's ex-anthropology students. Moore had learned that Hibben was looking for ibex. And because he had a lot of free time, he offered to help.

"I cabled back, 'Have at it,' " Hibben recalls, "and only a few weeks later Moore wired me that he had 18 Iranian ibex—can you imagine? It was spring, the time when ibex give birth, and the villagers simply went out and picked up the newborn."

Then a roadblock appeared. Major Moore had assumed that the animals could be flown out of the mountains on one of the daily Air Force transport planes. But the Pentagon said no. The animals must go by truck, over a one-lane, sandy and rocky camel track. It was 450 miles to Teheran.

Two trucks started, but only one arrived. Somewhere along that torturous road the shipment disappeared—truck, ibex, driver and all—presumably hijacked for the ibex fur and meat. Now only seven were left. They were flown to Hamburg for quarantine. Meanwhile, Hibben, frustrated and worried, had to sit home getting snatches of bad news by letter and an occasional expensive telephone call.

Finally, three months later, the seven ibex arrived in Albuquerque. Jack McDowell, a state game officer, remembers the day. "The ibex came out of the shipping box shy and gentle," he says. "They were the friendliest little fellows you ever saw. The little devils kept looking into my pockets."

Then out walked two animals that didn't look at all like the others. They were small and reddish, with a white saddle across their rumps, and they looked more like sheep than ibex. They *were* sheep—Elburz red sheep. The villagers evidently had pulled a fast one.

Within a week the two strangers—one male, one female—produced an offspring, "I was in the red sheep business without even trying," says Dr. Hibben, "and I was ecstatic; red sheep are so rare not a single other American zoo has one." Over the next year he watched while the male's horns grew large and handsome, curving down until the tips nearly touched his chin.

The Iranian ibex "herd," meanwhile, now had only four members, far too few to chance a release. Fortunately, the quartet settled down to captivity easily and quickly began multiplying.

With the sheep and ibex prospering, Dr. Hibben now sighted on the gemsbok, a large (up to 600 pounds), purplish South African antelope with astonishing, straight, four-foot horns. His face looks as if it's painted black and white with clown's make-up, and when he stares at you his odd, square ears stand straight out. His tail, tufted, seems to be stuck on like a misguided afterthought.

Once gemsbok ranged by the millions across the grasslands of southern Africa. But Bushmen, hunting with poisoned arrows and cruel, strangling-nooses, have reduced the number to only a few hundred.

Gemsbok would be ideal for the New Mexican wilderness, Hibben decided. They can go months without drinking, and they find desert shrubs succulent. As one ranch hand in South Africa said, "A gemsbok will thrive where a burro will starve to death."

"But the trouble was," says Hibben, "I knew of only one place where gemsbok were roaming free. That was in South West Africa's Gemsbok Park. But the government wouldn't let us capture any. In the meantime, of course, the Bushmen were killing the park animals off, so I couldn't quite see the government's position. But anyway, we had to look elsewhere."

For months he followed up rumors of gemsbok herds. But he located none until he wrote a South West African named Walter Schultz, who had a ranch near Okahandja, and who occasionally supplied zoos with wild animals. Schultz answered that he knew of a place in the dry wastes along the edge of the Kalahari desert where some scattered herds still roamed. So during another summer vacation period, Dr. Hibben flew back to Africa to try his hand at gemsbok capturing.

Just how do you catch a gemsbok alive? The ideal method, the men thought, would be with a tranquilizer gun, a rifle that shoots a hypodermic needle which injects a knockout substance when it strikes. "But we quickly found that gemsbok are extremely sensitive to such drugs," Dr. Hibben remembers. "The first two we shot died. We couldn't take a chance with any more of them."

The method they finally worked out was a wild and dangerous version of the cowboy's lasso technique. Hibben strapped a chair onto the front fender of a Land Rover. Then while one man drove through scrubland, another sat in the seat, secured by a safety belt, holding a pole with a noose on the end of it. Gemsbok run at about 60 mph, so the Land Rover must go faster. "It's a strenuous, exciting, frightening sport," says Hibben. "You have to separate one animal from the herd, get him running in a straight line, then lasso him with the loop. If you get a full-grown male, when he realizes he's caught he'll try to come right up on the car with you. And those horns are just like rapiers."

When a gemsbok was lassoed, the truck stopped, and 20 natives from a follow-up truck swarmed over the animal and forced him into a crate.

Once Dr. Hibben came close to losing his life. He was riding the catch seat, chasing three gemsbok, when suddenly a four-foot gully appeared in the path. Schultz, driving, slammed on the brakes. But he couldn't stop. "We must have slid in at 30 mph or so," says Hibben. "I thought my head was going to snap off. The radiator smashed into the fan and the whole works went up in steam."

A few days earlier, another catchman allowed the tip of his pole to touch the ground. The other end drove through his chest, killing him instantly.

Despite such dangers, 15 animals were caught, and six

females and two males were selected from them. Meantime, Dr. Hibben had been working with the government of South West Africa and the U.S. Department of Agriculture to designate Schultz's ranch as an official quarantine station. "For awhile there we thought we'd have to spend all that time and money—which we didn't have—shipping the animals to Hamburg, the nearest official station," says Hibben. "But then at the last minute our Senator Clinton Anderson, who had always been interested in the program, helped push the ruling along. We were safe."

Dr. Hibben had similar success with the kudu, a large African antelope which lives in rocky areas from the Sahara to the Cape of Good Hope.

Place a native of the mountain and desert regions of southern Africa in the dry terrain of southern New Mexico, and if he isn't a botanist, he may think he's home. The foliage looks the same, the temperature is the same, and today, in a mile-square fenced-in breeding area, the animals *are* the same. Here, progeny of those kudus, gemsbok and ibex painstakingly gathered abroad are thriving on the vegetation and climate, while biologists are making sure they're ecologically compatible.

Dr. Hibben estimates that the cost of the Barbary program has run to about $20,000 now. "And you know what? The state is coming out ahead," he says. "Tourists are flocking to the gorge to see the imports. They spend days rummaging around, and get a tremendous kick out of trying to photograph them. It takes a good man to spot a Barbary, and a fine photograph of one in the wild is still rare."

The Roy, New Mexico, Chamber of Commerce has counted out-of-state visitors, people who have come because of the sheep. "The figure topped 2,000 by last summer," says Hibben. "And if each of these people spent just $5, the sheep program already has paid for itself."

The whole import program has cost less than $100,000. "That may sound like a lot," declares Hibben, "but do you realize that in fish-and-game terms that much will build only a relatively small dam for a very small fishing lake? Why, last year the state spent half that on a study of the relationship between predators and wild turkeys."

Some of the people who were opposed are now vigorous supporters. Says Ladd Gordon, Director of the state Game and Fish Department: "Sure, there was apprehension. The local cattlemen were especially uneasy. But we've all pretty well got over our anxiety. Today I don't know of a single state conservation man or landowner who is still against the program."

Adds Governor David F. Cargo: "I think it's wonderful what they've done. You know, I heard a curious thing recently when I was talking to some visitors from Africa. One of them said how interesting it was that we were trying to propagate the animals about to become extinct, whereas in Africa just the opposite is happening. Actually, I think it's surprising we didn't start sooner."

"An earlier start would have helped," says Hibben, frowning. "Then there wouldn't be such a damn hurry. We're going after other species now—the eland, the springbok, various other antelopes and gazelles—but time is working against us. If we wait, some of these will be impossible to get."

"You know what I foresee?" he asks, his face brightening. "I'll bet eventually, when the Africans become more conservation-minded, they'll be visiting us, looking for animals to take back home for restocking."

The Creative Monkeys of Koshima

by Barbara Ford

A young macaque monkey on the little island of Koshima, Japan, dips a sand-covered potato in the sea and scrubs it briskly with its paw. When the potato skin is clean, the animal peels it off with its teeth and eats the potato.

Washing potatoes isn't ordinary behavior for a monkey. Monkeys in other areas, even nearby in Japan, don't wash their food. What makes the monkeys of Koshima do it?

The same group of monkeys, about 80 strong, follows another unique practice. They separate grains of wheat from sand by putting the sandy grains in water. The floating wheat is picked out and eaten. Who taught the Koshima monkeys this technique?

The evolution of a particular form of animal behavior isn't usually known, but the case of the Koshima monkeys is a little different. These animals and macaque monkeys in some 400 other colonies in Japan are among the most-studied animals in the world.

The Japanese macaque, a whiskery, medium-sized monkey with long, dense fur, is native to the country, but Japanese scientists didn't pay much attention to him until research on other monkeys around the world triggered their interest around 1950. Suddenly scientists realized that a prime subject for research was right in their own backyard.

Since that time, an extensive program of research on the animals has been undertaken with headquarters at Kyoto University. Results of research both in Japan and elsewhere are published in a scholarly journal called *Primates* (it's printed in English) and an elaborate Japanese Monkey Center has been opened to the public and researchers at Inuyama.

Visitors are also welcome at other Japanese monkey colonies, except where the location is too remote, as at Koshima.

Records are carefully kept at each colony on such data as the size of the group, the dominance order, births and deaths and any significant changes that take place. In smaller colonies like Koshima, almost every individual monkey is known, and many of them have been given names.

With records like this available, Japanese scientists can point out with some confidence that not long ago some inventive or "creative" female adult monkey on Koshima began washing the sand-covered potatoes left on the beach for the animals to eat. The potatoes are provided by a private society interested in maintaining the colonies.

Juvenile monkeys on Koshima watched the potato-washing female and imitated her actions. Soon a number of young monkeys could be seen industriously washing potatoes along the beach. No adult picked up the practice, however.

When the juveniles grow up, they'll continue washing potatoes, predict the scientists. They'll teach their young the practice and some day every monkey on Koshima will be washing potatoes and peeling them before they eat them. A cultural tradition will have been born.

Another creative female and the same sequence of events led, a little while later, to the grain separating behavior, the records indicate. Eventually this practice, too, will become a tradition.

These traditions in the making were filmed by an American scientist, Dr. R. C. Carpenter of Pennsylvania State University, and shown on television as part of a show entitled "The Evolution of Good and Evil." Dr. Carpenter shot the film, which will be available to scientists in Japan as part of the U.S.—Japan Cooperative Science Program.

In one of the sequences he filmed, a number of macaques are crouched on the Koshima seashore washing and peeling potatoes. An infant picks up and eats pieces of potato that his young mother drops for him.

"You can almost *see* the baby monkey learning how to wash potatoes," Dr. Carpenter says.

Another sequence he filmed shows a group of monkeys using different techniques to separate grains of wheat from sand. One throws the grains and sand in a stream, dashes downstream and picks up the floating grain. A more adept monkey holds grain and sand in his palm, immerses his hand in water and eats the grain.

An older female—too old, it seems, to have learned a technique—crouches downstream from a young monkey who is tossing grain and sand in the stream. The young monkey gets most of the floating grains but a few escape downstream, where the female grabs them.

Since grain separating is a newer practice, not quite as many monkeys are seen participating in it as in potato washing. And they seem, as Dr. Carpenter notes, less involved in the practice and less excited about it.

What prompted the creative monkeys of Koshima to wash potatoes and separate grain?

Dr. Carpenter thinks they began the practice simply in order to survive. Although monkeys have lived on Koshima for about 300 years, the tiny, windswept bit of land doesn't offer its primate population an adequate diet. The macaques on Koshima are noticeably smaller than the macaques in other areas of Japan. The amount of potatoes and grain brought in from the mainland to supplement the animal's diet doesn't quite make up for the lack of food on the island.

It was to make as much food as possible available, then, that the creative monkeys began cleaning the potatoes and grain that are simply dumped on the sandy beach, Dr. Carpenter theorizes. The cleaning not only increases the food supply, it protects the animals' teeth, he adds.

In a much larger monkey colony called Takasakyami on the big island of Kyushu, the macaques also are fed potatoes, but the vegetables are not covered with sand. Sometimes a Takasakyami macaque will dip a potato in water, apparently to coat it with salt, but he will not wash it as a Koshima monkey does. Would dirtier potatoes make a difference?

No one knows yet, but to test the idea there are plans to offer sand-covered potatoes to the Takasakyami monkeys. If there are any creative individuals among the colony, the results should be interesting.

Shark! Overrated Demon or Genuine Scourge?

by William and Ellen Hartley

In October, 1944, a large American magazine published an astonishing article entitled, "The Shark is a Sissy." Much of it was based on a navy pamphlet, "Shark Sense," which noted that, "There is practically no danger that an unwounded man floating in a life jacket will be attacked by a shark."

That same year, Lt. Com. H. R. Kabat and E. S. Hahn wrote and published an account of Kabat's battle with a shark off Guadalcanal. Unwounded and in a life jacket, the lieutenant commander had been savagely mauled until a ship came to his rescue. Nearer home, there were at least two recorded shark attacks off the Florida coast and one in the Caribbean. A year earlier, in a particularly grim incident, parts of the body of Clyde Kelly Ormond, Jr., had been found in a Tiger shark captured off Miami Beach, Florida.

The Navy pamphlet, of course, was designed to improve morale among war pilots and seamen who feared sharks more than enemy fire. Unfortunately, it inspired a body of comforting belief in the harmless shark idea. More unfortunately, skindivers recently have delighted in perpetuating the theory that "if you ignore them they'll go away."

Not all skindivers go along with the theory, however. On Dec. 8, 1963, a 23-year-old insurance salesman named Rodney Fox was competing with some 40 other divers in the annual

South Australian State Skindiving and Spearfishing Championship. While diving near Aldinga Beach south of Adelaide, he suddenly found himself literally propelled through the water by a tremendous, crushing force. A huge shark, called a White Pointer by Australians, had closed its jaw around the diver's chest and back, and was rushing away with him.

Fox managed to tear himself free. When the creature attacked again, the diver clutched it like a groggy boxer and was carried to the bottom. Releasing his grip, Fox then struggled to the surface and saw with horror that the water was stained with his blood.

In its final assault, the shark suddenly veered away, seizing Fox's fish float on which speared fish were stored. Since Fox was tied to the float by a long rope, he was dragged through the water until the line broke. Friends saved him, but his injuries were frightful. His chest, back, left shoulder and side had been torn wide open, his right hand and arm slashed to bone; and pieces of flesh had been ripped from his body. Almost a year passed before he regained his health.

At that, Fox was unbelievably lucky. On August 15, 1959, an Army lieutenant named James C. Neal was scuba diving off Panama City, Florida. When he failed to surface, friends went for help. Later, bits of Neal's clothing and diving equipment were found with unmistakeable marks of shark teeth.

The hard fact is that the supposedly harmless behavior of sharks has *never* jibed with reported facts. Today, most people are aware that these creatures can be formidably dangerous. And during recent years, scientists have been hitting at the core of the problem—the shark itself—to examine the unique sensory equipment that enables the animal to locate its prey in the water.

Modern shark research was largely stimulated by a New Orleans Shark Conference in April, 1958, when scientists were amazed by the limited knowledge of shark behavior. They established the Shark Research Panel of the American Institute of

Biological Sciences to serve as a clearing house of shark information. For 1966, its Shark Attack File reports 41 shark "incidents" in which nine persons were killed. Eleven attacks took place in North American waters. None was fatal; but during the past 50 years, about 70 deaths have resulted from more than 200 known attacks in our part of the world. Australian and South African waters appear to be most dangerous, with United States and Latin American coastal waters running second.

The Shark Attack File also lists sea and air disasters in which victims were either mutilated or eaten by sharks after death, or were killed as a result of shark assaults. In 1966, the total was 281 persons. This was the worst disaster toll for many years, and a clear indication that shark research is imperative. For too little is really known about their habits and behavior. Shark repellents developed during the war years were never satisfactory and a fully successful preparation to ward off possible attacks still remains to be found. A compressed air bubble screen—a kind of shark fence—was tested during 1961 by Dr. Perry W. Gilbert at the Lerner Marine Laboratory of the American Museum of Natural History in Bimini, Bahamas. Tiger sharks, the test animals, showed no objection to the bubble bath.

A new kind of shark research has been under way within the past four or five years, however, at places such as the Lerner Laboratory; the Marine Biology Laboratory at Woods Hole, Mass.; the University of Hawaii; the Cape Haze Marine Laboratory where Dr. Eugenie Clark has been studying shark vision; and the Institute of Marine Science, University of Miami, Florida. (South Africa and Australia have also been active in shark research).

New work in Miami, started by Dr. Warren J. Wisby and carried on by Dr. Arthur A. Myrberg and others, is particularly interesting because it approaches the shark menace where it begins—in the creature's sensory systems. How do sharks hear and see or feel; *what* do they hear and see and feel?

Dr. Myrberg explains: "There we're many reports from skindivers and fishermen that they never saw sharks around until they had speared a fish. Then a shark would arrive. Oddly, they often came from upstream, so they couldn't have smelled the fish or its blood. This suggested that sharks were not only capable of hearing sounds, but they were also able to locate sounds in three dimensional space."

But how?

Human hearing ranges from about 16,000 cycles per second down to about 40-100 cps—high to low. Studies of lemon sharks, chosen because they are excellent laboratory animals, rather common and quite dangerous, showed that their hearing range was far below human hearing (from about 1,000 cps down to 7.5 cps). Their most sensitive area was approximately 40 cps. This clue meant that if sharks were using sound for locating prey, they might be attracted primarily by sounds in the low range.

It happens that low sounds travel a tremendous distance in water. What's more, sounds of low frequency are those brought about when there is turbulence in the water—a fish fluttering after being speared, a swimmer thrashing around.

To determine what sort of sounds a speared fish made, the experimenters first went to the reefs outside of Miami. Using recording equipment, one diver would place a hydrophone near a cave occupied by a fish. When a second diver speared the fish, a recording operator in a boat would begin recording.

The sounds were then analyzed and reproduced. Next, the experimenters went into shark-inhabited waters. After spotting a shark from the air, they would bring a boat to the area, hang a transducer over the side, and produce three kinds of sound—low frequency pulsed, high frequency pulsed, and low frequency continuous sound.

Results were amazing. Eighteen sharks of several species turned and *headed directly toward the boat* when the low frequency pulsed sound was played—essentially, the sound made by a struggling fish. The animals not only became interested in hearing the sound, but also could locate that sound at a distance of at least 600 feet.

So far so good, but here the plot thickens. A shark has ears, but because of the speed of sound in water—about five times faster than in air—the shark's hearing organs are not spaced nearly far enough apart to permit any sort of "hearing triangulation." Nor do sharks have a gas bladder like fish—a sort of resonator which is probably used to amplify sounds.

The best bet seemed to be the "lateralis," or lateral line system, which runs along the side of a shark's body. This consists of small sensory cells called neuromasts, which join in a main nerve channel entering the brain near the ear. It has long been known that the shark's lateralis system is responsive to low frequency vibrations, but no one had fathomed its significance.

By destroying the inner ear in laboratory experiments, it was found that the sharks were still aware of the lower range of frequencies. Thus the lateral line system *alone* appeared capable of giving information to the animal. How? Scientists suspected that the unique sensory system might be stimulated by pressure or displacement. Pressure is simply physical force. Displacement means that something has moved—in this particular case, water particles. A displacement measuring device soon revealed that the shark responds to displacement. Pressure was not the important factor. A more sensitive device for measing displacement may soon establish the distance from which sharks can receive such signals or clues. To you and me, what this means is that thrashing around in the water should be avoided at all costs unless we want to attract sharks.

The question of shark's vision has been a subject of considerable controversy. Most authorities have agreed that sharks have low visual acuity but reasonably good equipment for distinguishing an object against its background.

Laymen have sometimes advanced the notion that sharks see poorly in murky water or at night. But Rodney Fox was attacked in murky water. So was 13-year-old James McKee of Marathon, Florida. His knee was bitten on March 29, 1959. And Barry Wilson was killed by a shark on Dec. 7, 1952, while swimming in murky water at Monterey Bay, California.

Thor Heyerdahl, author of "Kon-Tiki," states in his book that most of the sharks he saw while crossing the Pacific attacked their prey toward twilight. Robert Walker, after capsizing off the Florida coast in September, 1959, was attacked repeatedly at night while drifting with a boat cushion for support. Walker, in fact, noted that the attacks began with nightfall. He was rescued, but almost lost his life from gashes that required 60 stitches to close.

Obviously, a swimmer can't see well in murky water. How well can sharks see under similar conditions? And can a shark see color? This is tremendously important for survival. If particular colors attract them, they're good colors to avoid in the water.

For a long time, scientists could find no cone cells in the eyes of sharks they dissected (Cone cells are responsible for color vision; "rod" cells convey black and white information). This suggested that sharks could not see color, but responded only to brightness.

But British and other scientists observed that when a shark attacked a grey lifeboat with an area painted international orange, the attack seemed to concentrate on the color spot. Moreover, recent studies of the shark's eye indicate that there *are* cones. They also reveal that vision in some sharks is not as poor as many investigators once thought. Their range of adaptability to darkness is roughly equivalent to that of a human being. This is a good reason not to swim at night or in murky water. Besides his dark adaptation, the animal also has the advantage of hearing and smelling underwater.

Despite discovery of cone cells in shark eyes, tests did not prove the presence of a color sensitive system, but neither did they disprove it. Some scientists posed an interesting question:

Does the shark have a second system for seeing color?

One Miami experimenter, studying the shark's "critical flicker fusion" (the point above which an eye no longer sees a flickering light) turned up the fact that the visual mechanism of the shark is indeed composed of two systems. This simply means that color perception in the shark is highly probable. As yet, however, we don't know *what* colors sharks see, or exactly how the brightness factor enters in. At Cape Haze, Dr. Eugenie Clark trained a lemon shark to tell the difference between white and red targets. Did it really see the colors? Dr. Clark thinks that it responded to brightness.

Today, shark behavior is under the microscope. "Few scientists know much about what sharks do," Dr. Myrberg says. "Few have information on patterns of reproduction, attack, feeding, and so forth . . . This area of study will be of increasing importance."

To the family vacationing on a beach this summer, what sharks do is more important than why they do it. Recent research has taught us new things, confirmed past speculations and exploded many of the old myths. For example, Dr. Leonard P. Schultz, curator in charge of the Division of Fishes, U.S. National Museum, identifies some 27 out of 250 shark species as being dangerous. But some of these 27 have been rumored to be "harmless" even by people who should know better.

The bull shark, common to Florida waters, has been regarded by many divers as a harmless bluffer.

But evidence suggests that bull sharks and their close relatives are far from harmless. This is also true of the nurse shark, abundant in Florida waters and found as far north as lower New England. The nurse often swims away if disturbed. But when Johnny Bowers grabbed a five-foot nurse shark by the tail, while swimming off Miami Beach on April 30, 1958, the creature twisted around and seized Bowers' leg. A year later, a 13-year-old boy was grabbed by a nurse shark near Delray Beach, Florida. The shark, a little two-and-a-half footer, hung tenaciously to the boy's arm while the victim swam to land.

We now know that sharks can hear surprisingly well, are attracted by splashing, and are inclined to investigate hungrily. Ten years ago, writers were still suggesting that the best way to scare sharks was to thrash wildly in the water.

We have known for a long time that sharks can smell blood or juices from wounds in fish or men. But we didn't realize until recently that their vision is reasonably good, and that they can adjust well to seeing (as well as hearing and smelling) in gloom or dark. A full list of safety suggestions, based in part on Dr. Perry Gilbert's advice in the book, "Sharks and Survival," appears on the opposite page.

Scientists still want to know more about what colors attract sharks. And there is grave need for a fully effective shark repellent. The need was dramatized horribly on May 16, 1966, when Luceretio Nazareno, radio operator on the ship *Pioneer Cebu*, staggered to the deck after tapping out a final message: "Sinking, sinking, sinking." The ship had been caught in a typhoon off the island of Cebu in the Philippines.

Nazareno and some other survivors were reported safe at Cebu City several days later. But the captain and 135 of the 262 *Pioneer Cebu* passengers were lost. The world "Shark Attack File" lists all of them as victims of sharks. It could easily happen in North American waters.

Most dangerous sharks in North American waters

Name	Color	Full size	Main locale
Bay shark	Yellow-brown to grey	to 12 feet	lower Pacific coast
Blue shark	blue with white under	12 to 20 feet	Pacific coast; Atlantic; tropical waters
Bull shark	grey	to 12 feet	Atlantic coast
Dusky shark	grey	to 12 feet	Atlantic coast
Galapagos shark	grey-brown	to 12 feet	lower Pacific coast; Atlantic
Great Hammerhead	grey-brown		Atlantic coast
(Scalloped Hammerheads are found on both coasts, as are Common or Smooth Hammerheads. All dangerous. Head is unmistakable.)			
Lemon shark	yellowish	to 11 feet	New York to Florida
Atlantic Mako	blue-grey with white under	to 12 feet	Warm Atlantic
Nurse shark	brownish-yellow	to 14 feet	Atlantic and Pacific
Tiger shark	barred grey	to 15 feet or more	Atlantic
White shark* ("Man-Killer")	grey to black with white under	to 20 or more feet	Atlantic and Pacific

*Most dangerous, but not numerous. In terms of abundance, the Lemon, Tiger, , Bull, Hammerhead and Nurse sharks represent the greatest danger. The huge Whale shark (up to 60 feet) is harmless; so is the big Basking shark. There appear to be no recorded attacks by the common Atlantic Sand shark.

Life Among the Gorillas

by George B. Schaller, Ph.D.

I entered the office of John T. Emlen, professor of zoology at the University of Wisconsin, one day in January, 1957, to ask a question. Doc leaned back in his chair: "Would you like to study gorillas?" I was working in bird behavior at the time, and took the question lightly.

"Sure," I replied. Two years later that impulsive answer had taken me from an office in Wisconsin to the forests of Africa.

My wife, Kay, and I arrived in Rumangabo, the headquarters of Albert National Park, on February 14, 1959. From then, through 1960, we stalked, lived with and studied—in their natural environment—one of the world's most interesting creatures. But some of our most vivid experiences with these shy, shaggy giants occurred during those first summer months.

Our working camp, a 25-year-old hut, was located on a saddle between Mts. Mikeno and Karisimbi near a site known to the natives as Kabara (the "resting place").

On one of our first mornings there, while Kay put the cabin in order, I headed north over the rolling terrain toward Bshitsi. My path led through dense stands of lobelias. Slightly ahead and to one side I suddenly heard a quarrelsome, high-pitched scream. I crept ahead and from the cover of a tree trunk looked out over a shallow valley. A female gorilla emerged from the vegetation and slowly ascended a stump, a stalk of wild celery casually hanging from the corner of her mouth like a cigar. She sat down and holding the stem in both hands bit off the tough outer bark, leaving only the juicy center which she ate.

Another female ambled up with a small infant clinging to her back. She grabbed a stalk of wild celery near the base and pulled it up with a jerk. She then pushed a swath of vegetation down with one hand, squatted and ate, scattering the strips of celery bark over her lap.

To obtain a better view I became incautious and let myself be seen by one female. She emitted one short scream and ran off into the undergrowth. A large youngster, weighing about 80 pounds, climbed up the sloping trunk of a tree, looking in my direction intently, and descended rapidly. Suddenly seven animals, with a large silverbacked male bringing up the rear, filed by 100 feet from me. He paused briefly, peering at me from the cover of a screen of herbs, only the top of his head showing. After a harsh staccato of grunts, which apparently functioned as a warning to me as well as to the group, he hurried away, closely followed by three females and four youngsters.

Group I, as I referred to the first gorillas I had seen in this area, was not the only group around Bshitsi at that time. Later, I heard a male beat his chest about 100 yards from the animals I was watching, and the following morning inspection of the forest revealed that no less than three groups had nested close to each other. One was large, comprising 19 animals, but the third group contained only five. These three gorilla bands remained in the same part of the forest for five days.

Most of my visits to a group began at the site where I had last seen it the previous day. I followed the trails of the gorillas through the trampled vegetation, never certain if the animals had gone 100 yards, a mile, or doubled back. The trails always had an interesting story to tell, and I enjoyed my saunterings as much as the sight of the apes themselves. When gorillas were feeding, they fanned out, leaving many trails littered with discarded celery bark and other food remnants. When they were traveling, they moved in single or double file, only to rest after awhile close together on an open slope. Sometimes a musty odor, like that of a barnyard, permeated the air, and I knew that it was the site where the animals had slept the night before.

Somewhere not far ahead in the undergrowth were the gorillas, often without a sound to reveal their presence. To track them over this last piece of trail, never certain when a shaggy head would rear above the vegetation, never certain that an attack by the male would not follow careless approach, was the most tense and exciting part of the day. Cautiously, I would take one, two, three steps, all senses alert, listening for the snapping of a branch or the rumbling of a stomach.

Frequently my first intimation that the gorillas were near was the sudden swaying of a plant, jarred by a passing animal. Then there were two courses open to me: I could hide and watch, or I could remain in the open with the hope that, over days and weeks, they'd become accustomed to seeing me near them. I usually walked slowly and in full view toward them and climbed a stump or tree branch where I settled myself without paying obvious attention to the animals. By choosing a prominent observation post, I was not only able to see the gorillas in the undergrowth, but they could keep an eye on me.

Animals are far more accurate interpreters of gestures than man. It is not easy for man to shed his arrogance before an animal, to approach it in utter humility with the knowledge of being in many ways inferior. Casual actions are often sufficient to alert gorillas and to make them uneasy. For example, a direct unwavering stare is a form of threat. Even while watching gorillas from a distance I had to be careful not to look at them too long without averting my head, for they became uneasy.

Establishing rapport was fairly easy. Gorilla senses are comparable to those of man. The apes are quick at spotting slight movements, and often they watched my approach before I was even aware of their presence. Hearing, too, is well developed in gorillas, but the sense of smell seems to be relatively poor.

On the last day of August, as N'sekanabo, a park guard, and I clambered up the boulder-strewn depths of Kanyamagufa Canyon, a tremendous roar filled the chasm. We peered up at a silverbacked male who, surrounded by his group, stood motionless at the canyon rim looking down at us. As unobtrusively as possible we retraced our steps under the watchful eye of the male, feeling like chastised children for having invaded his domain. We climbed up the opposite wall of the canyon, where I was able to see the gorillas well. It was Group IV. No other group taught me as much or took a greater hold on my affection.

On September 4, I came upon these gorillas feeding slowly on a steep slope about 100 yards above me. I sat at the base of a tree and scanned the slope, trying to pinpoint the whereabouts of the four silverbacked males in the group. Large male gorillas are the most alert, unpredictable, and excitable members of a group and hence the most dangerous. Squatting with his back toward me was "Big Daddy," easily recognizable by the two bright silver spots on his gray back. As he turned to rest on his belly, he saw me, gave me an intent look, and emitted two sharp grunts. Several females and youngsters glanced in his direction, then ambled to his side. Big Daddy was the undisputed leader, a benign dictator who by his actions determined the behavior of the other animals. He stood now looking down at me with slightly parted lips, his mighty arms propped on a knoll, completely certain of his status and his power, a picture of sublime dignity.

"Daddy Junior" was the striving executive type, second in command. In matters such as determining the direction of travel and the time and duration of rest periods, the females and youngsters ignored him. He lay by himself on his back, one arm slung across his face, oblivious to the world.

"The Outsider" roamed around the group, intent on his own doings. He was a gigantic male in the prime of life. His nostrils were set like two black coals in his face, and his expression conveyed an independence of spirit and a glowering temper. His gait was rolling like that of a seaman. To estimate accurately the weight of gorillas in the wild is difficult, but I believe that the

Outsider must have weighed 450 to 480 pounds. Gorilla males are often said to weigh 600 pounds or more, but these are obese zoo animals.

The fourth silverback was "Splitnose," so named for the ragged cut that divided the upper part of his left nostril. He was young, his back barely turned silver, and he lacked the quiet sureness of action which characterized the other adult males. As if to compensate for his uncertainty of mind, he was highly vociferous whenever he saw me, roaring again and again. But none of the other animals responded visibly.

Apparently D. J. had hatched a plan, for suddenly he left his resting place and circled uphill. Then, stealthily, he angled toward me, keeping behind a screen of shrubs. To orient himself he stood up occasionally to glance over the vegetation. As soon as I looked directly at him, he ducked and sat quietly before continuing his stalk. He advanced to within 30 feet of me before emitting a terrific roar and beating his chest. Before the echo had died away, he peered out as if to see how I had responded to his commotion. Never was I able to get used to the roar of a silver-backed male. The suddenness of the sound, the shattering volume, invariably made me want to run. But I derived immense satisfaction from noting that the other gorillas in the group startled to a roar just as visibly as I did.

With the male so near, I became uneasy. Cautiously, I ascended a tree. One of the ten females in the group left Big Daddy and ambled to within 70 feet of me to sit on a stump, her chin propped on folded arms. Slowly, the whole group advanced toward my tree. I felt a spasm of panic, for the apes had never behaved in this manner before. They congregated behind some bushes. Three females carried infants and two juveniles ascended a tree and tried to obtain a better view of me through the interlacing vines. In the ensuing minutes we played a game of peekaboo: whenever I craned my neck to see the gorillas more clearly, they ducked their heads, only to pop forth again as soon as I looked away. One juvenile, perhaps four years old, climbed into a small tree adjacent to mine, and there we sat, 15 feet apart, each nervously glancing at the other.

Junior, the blackbacked male in the group, stepped out from behind the shrubbery and advanced to within ten feet of the base of my tree, biting off and eating a tender leaf of blackberry bush on the way. He stood on all fours and looked up at me, mouth slightly open. There was recklessness in his face and a natural mischievousness. At the same time his look conveyed a critical aloofness as if he were taking my measure, not quite sure if I could be trusted.

He was the only gorilla who seemed to derive any sort of satisfaction from being near me. For weeks, he would leave the group to sit by me, either quietly watching my every action or sleeping with his back toward me. Today he was still somewhat uncertain, as his indrawn lips showed. Man too bites his lips when nervous. Occasionally he slapped the ground with a wild overhand swipe, using the palm of his hand, then slyly looked up at me, apparently with the hope that his wanton gesture had been startling. The other members of the group rested quietly. Every 15 or 20 minutes one of the males jerked out of his slumber to roar once or twice before reclining again to continue his nap.

All apprehension had long since left me. Not once had the actions of the gorillas portrayed ferocity or even outright anger. The silver-backed males were somewhat annoyed, to be sure, and several animals were excited, but all this was offset by their curiosity and rapid acceptance of me. As long as I remained quiet, they felt so safe that they maintained their routine even to the extent of taking naps beside the tree in which I sat.

During ensuing days Group IV traveled through several deep ravines parallel to the slope of Mt. Mikeno in the direction of Kabara. The gorillas had become used to my presence: the females hardly responded at all, and Big Daddy did little more than grunt briefly in annoyance each time he first saw me.

On several occasions Junior approached within 60 feet or less and shook his head from side to side at me. It was an odd gesture; one that seemed to signify "I mean no harm". To see what gorillas would do if I shook my head at them, I waited until Junior was 30 feet away, paying close attention as I rewound the film in my camera. Then I began to shake my head. He immediately averted his face, perhaps thinking that I had mistaken his steady gaze for threat. Then, when I in turn stared at him, he shook his head. We continued this for ten minutes. Later, when I inadvertently met gorillas at close range, I employed head shaking as a means of reassuring them. They seemed to understand my good intentions.

The animals generally bedded down for the night at dusk and began to stir one hour after sunrise, having slept some 13 hours. They were silent at night, except for the rumbling of stomachs or breaking wind; I never heard them snore. When excited, a male sometimes beat his chest during the night.

Group IV moved down into the zone of bamboo during the latter part of September, where in the maze of stems I was unable to observe them well. When a few days later they crossed Kanyamagufa Canyon to reappear in their old haunts on the slopes of Mt. Mikeno, one of the females carried a newborn infant. Carefully she held the helpless creature to her chest, supporting it with one arm. Newborn gorillas are tiny, weighing only four or five pounds, and they are so weak that they are unable to hold on to the hair of their mothers for more than a few seconds. Their movements appear disoriented, and they have a vacant look, just like a human infant. At the age of one month, young gorillas begin to follow with their eyes the movements of other members of the group. By the age of two and a half months, infants show a marked increase in movement. Their upper and lower middle incisors have appeared, and the youngsters now reach for and chew on branches and vines. For the first time the mother may place her offspring on the ground by her side and watch over it carefully as it shakily tries to crawl. The rate of development of a gorilla infant is roughly twice as fast as that of a human baby. I suspect that few wild gorillas grow older than 30 years.

Silverbacked males dominate the group. Similarly, females dominate juveniles, and juveniles dominate infants that stray from their mothers. Once, at the beginning of a downpour, a juvenile sought shelter beneath the leaning bole of a tree. But when a female hurried toward the tree, the juvenile vacated its seat and fled into the rain. As soon as the female had taken over the dry spot, a silverbacked male emerged from the undergrowth. He sat down and with one hand pushed the female gently but determinedly until she was out in the rain, a victim of the rank system.

Unlike silverbacked males in a group, females seem to lack a definite and stable rank order among themselves. It is perhaps significant that quarreling erupts mainly among the females, with silverbacked males taking no active part.

Late in September I spent many delightful hours in the company of Group II. There were 19 gorillas including the silverbacked male, three blackbacked males, six females, five juveniles and four infants. This group contained an old female with sagging, wrinkled breasts and with kindly tolerant eyes. She lacked an infant of her own, but her affection found an outlet in Max. Max was an impish and rambunctious six-month-old infant who could never sit still. When his mother held him, he poked at her eyes with his fingers until she averted her head. He stiffened and arched his back, and, as soon as his mother loosened her hold, he twisted and wriggled until finally she placed him on the ground beside her. Then, as often as not, he stood by her side, hands raised above his head, wanting to be picked up again. When this gesture was ignored, he bumbled off. If the old female was nearby, he hurried into her arms. One sunny morning, as the female slept on her belly, Max climbed up on her back, walked forward and stood on her head, slid off, and then, as she rolled over, climbed up on her abdomen. She grasped Max and held him against her belly with one hand. His mouth was partially open, with the corners pulled far back into a smile. The female loosened her hand, and Max grabbed it and gnawed at her fingers. They then toyed with him, touching him here and there as he attempted to catch her elusive hand. Finally Max lay on his back on her belly, waving his arms and legs with wild abandon, and the old female watched the uninhibited youngster with obvious enjoyment. Suddenly Max sat up and, with arms thrown over his head, dove backward into the weeds.

For days I watched Max cavort—with almost as much enjoyment as the old female gorilla. At the end of September, I assessed the results of my work at Kabara. I had found five gorilla groups and watched their behavior for over 50 hours, learning many new things about the life of this ape. The thing that had impressed me most was that the gorillas had proved amiable far beyond my wildest expectations. I'd learned to track them by their spoor and to stay in close proximity with them. All told, it was a rare experience.

The Elephant Seals of Ano Nuevo and Loser's Beach

by John M. Leighty Jr.

Right now, strange three-ton animals are dragging themselves onto the tiny island of Ano Nuevo off the coast of California.

These creatures are bull elephant seals. They spend nine months of every year in the open sea, but return to Ano Nuevo in December or early January for three months.

They come for two purposes: to fight and to copulate. The two things, according to Burney J. LeBoeuf, are irrevocably intertwined. LeBoeuf, associate professor of psychology and biology at the University of California at Santa Cruz, explains that males who wish to mate must arrive about two weeks before the females come ashore and battle it out with the other bulls.

Only a handful of males will win enough fights to achieve high enough status in the rigid elephant seal hierarchy to gain mating rights. The others are doomed to a long, dry winter.

The odds are tremendous, according to LeBoeuf, who has studied elephant seals for seven years and is probably the world's foremost authority on the sexual-social hierarchy of the super seals. Of the 150 bulls who show up at Ano Nuevo, only the top ten or twelve will achieve mating privileges.

The males begin fighting for dominance in early December. Oftentimes the fights are no more than a show of force. A bull will make a threatening noise by lifting his head back and thrusting his long, flexible snout—called a proboscis—into his mouth and blowing. Burps of scraping sound emerge that can be heard for three miles.

But if that doesn't frighten the opponent away, a bloody battle ensues. The two- to three-ton, 16-foot animals take to the water to improve their mobility (they hit speeds in excess of two mph on land) and lunge at each other with the upper parts of their bodies.

Striking at each other with long, sharp canines, they deal blows so powerful to their opponent's back and neck that the blubber can be seen to shake from the force. Two bulls may fight for as long as 45 minutes and color the water red with their blood.

"No elephant seal has ever been known to die in combat," says LeBoeuf, who explains that while elephant seal battles are not fatal, the male's chest becomes hard and calloused and is the animal's badge of courage.

The seal which retreats first automatically loses and from that time on will move out of the winner's way whenever they meet. The winner doesn't have it easy either. He may in turn be defeated by another bull. The final result of all these individual battles, explains LeBoeuf, is a strict social order.

And it is not an egalitarian one. The losers in any one season, and this may include up to 90 percent of the male population, stand little chance of mating. Some males, in fact, go through life without copulating. And at Ano Nuevo, the subordinate seals will leave the island altogether, swimming to the Santa Cruz shoreline to claim a plot of land along an area nicknamed, appropriately, Loser's Beach.

But for the minority of dominant bulls, the rewards are tremendous. "It's a very extreme sort of thing," says LeBoeuf of the males' harems. Out of the original 150 males, the six top fighters claim 90 percent of the females, or about 40 apiece. "The other ten percent," says LeBoeuf, "are maybe divided among six other males."

The top male, called the alpha bull, gets the most sex of

anybody. He must be a potent lover as well as a great fighter. He often copulates twice within five or six minutes and mates with as many different cows as possible. The alpha bull (and some of the other dominant males) will go the entire 90-day mating season without eating. During this time he sleeps only when the other males sleep and must repel any bulls who encroach on his harem.

After the rookery has been staked out by the strongest bulls, the cows begin arriving. They are pregnant from the previous year's mating and give birth to pups within 34 days. They are then re-impregnated. About the end of March the entire herd gathers and swims away.

Although nearly impossible to study at sea, LeBoeuf says watching them on land is more important since most of their biological processes—copulation, birth, etc.—take place out of water.

"They're animals you can mark and follow for a lifetime," says LeBoeuf, who keeps track of who's who by bleaching on their hides such names as Max, Harry, George, Nip, Tuck, Spiro and Superseal. "What you do is simply wait until they go to sleep and then sneak up on them and literally mark anything you want to on their backs."

A characteristic of the species is a lack of fear of man. It is possible to sit on a bull without awakening it, although it would charge if angered. Even then, they aren't dangerous because of their slow speed. "They won't chase you more than ten feet," says LeBoeuf. "If you trip, you're in trouble."

During his investigations LeBoeuf has discovered that the language used by males differs among the various colonies along the California coast. Elephant seals return to the same rookery each year and a distinct difference in vocalization, consisting of a series of burps or honks, can be detected between one rookery and another.

"At Ano Nuevo, the burps come at about one per second," says LeBoeuf. "This occurs slower or faster at other rookeries. It's quite different. You can't miss it.

"It's pretty clear this is a learned component and not inherent," LeBoeuf says of the language pattern. "This is the only mammal besides man where there is a resemblance of dialect."

Female elephant seals show no such distinct language pattern.

The giant mammals, which live to be about 40 years old, were once almost wiped out by whale hunters who sought their blubber. In 1892 a lone surviving colony was rediscovered on Guadalupe Island, 150 miles off the Baja California coast. They were put under protection of law and proliferated. In 1972 they were removed from the endangered species list.

According to LeBoeuf, there are now upwards of 40,000 elephant seals. They live mostly on eight rookeries stretching from Guadalupe to the Farallon Islands near San Francisco.

Postscript

In February, 1975, the first elephant seal pup was born on Loser's Beach to a lone, stray female. In 1976, however, eight females took up residence on the beach and gave birth to pups, thus establishing rookeries on the mainland for the first time. Park rangers now guard the area, allowing visitors to watch the mammals from above the beach area. Tours to the viewing area are on a first come basis and on some weekends up to 2,500 persons are turned away.

PHOTOGRAPHY CREDITS

Courtesy of:
The American Museum of Natural History
John M. Leighty, Jr.
San Diego Zoo Photo
Science Digest Magazine
U.S. Fish and Wildlife Service

Book design by Sheila Lynch
Jacket and cover by Preston Fetty

Editorial Coordinator: Cheryl Johnson
Assistant Coordinator: Nancy Jessup